The Rhesus Monkey *Macaca mulatta*. (From Wikimedia Commons, Maria Cartas, http://species.wikimedia.org/w/index.php? title=Macaca_mulatta&oldid=1756192.)

Erythrocytes of the Rhesus and Cynomolgus Monkeys

Erythrocytes of the Rhesus and Cynomolgus Monkeys

Chester A. Glomski, M.D., Ph.D.
Department of Pathology and Anatomical Sciences
Jacobs School of Medicine and Biomedical Sciences
University at Buffalo
The State University of New York

Alessandra Pica, B.Sc.
Department of Biology
School of Polytechnic and Basic Sciences
University of Naples Federico II, Italy

Jessica F. Greene, Ph.D.
Greene Consulting
Portland, OR. U.S.A.

CRC Press
Taylor & Francis Group
Boca Raton London New York

CRC Press is an imprint of the
Taylor & Francis Group, an **informa** business

Front Cover: Image of Rhesus Macaque (*Macaca mulatta*) with two babies in Shimla Himachal Pradesh near the Jakhu Temple. Photo copyright Jan Tolkiehn. Used with permission. All rights reserved.

CRC Press
Taylor & Francis Group
6000 Broken Sound Parkway NW, Suite 300
Boca Raton, FL 33487-2742

First issued in paperback 2019

© 2016 by Taylor & Francis Group, LLC
CRC Press is an imprint of Taylor & Francis Group, an Informa business

No claim to original U.S. Government works

ISBN-13: 978-1-4987-3359-5 (hbk)
ISBN-13: 978-0-367-37703-8 (pbk)

**Visit the Taylor & Francis Web site at
http://www.taylorandfrancis.com**

**and the CRC Press Web site at
http://www.crcpress.com**

Contents

Preface

The erythrocyte is perhaps the world's best-known cell, a premise consistent with the French expression *Le sang c'est la vie* (i.e., "Blood is life"). The red cell made its debut in a few scattered invertebrates, and thereafter ventured upon a conceptual phylogenetic odyssey to the classes of cold-blooded vertebrates (poikilotherms), then onward to the avians (the first homeotherms), and thereafter to the mammals, which include nonhuman primates and man.

The intent of this book is to address the morphologic, quantitative, and generative aspects of the erythrocytes of the rhesus monkey *Macaca mulatta* and the cynomolgus monkey *Macaca fascicularis* (long-tailed macaque, crab-eating monkey). These two species are the most commonly selected nonhuman primates for basic science and clinical medical investigations. The hemopoietic cells of man and the rhesus monkey display an intimate homogeneity. Their functional activities are close and at times identical (e.g., the CD34+ marker identifies the hemopoietic stem cell of both man and the rhesus monkey). The cynomolgus monkey was enlisted in biomedical studies at a time when rhesus monkeys were not available in sufficient quantities. It has gained increased use in the Far East and in the Western world. It is, for example, employed in the current development of a vaccine against the deadly Ebola virus.

The in-depth studies of the erythrocytes of the rhesus and cynomolgus monkeys are part of the overall scientific examination of mammalian hemic cytology. The immediate application of these studies to human cellular biology and the investigation of human medical problems make these investigations even more vital.

The areas that this text embrace include the erythropoietic profiles of normal and abnormal macaques of both sexes and of all age groups as investigated using contemporary electronic methodology. Other topics include the role of stress as it is perceived by the monkey and how it impacts erythrocellular values; the training of the monkey to be a cooperative, unperturbed subject for hematologic study; the role of medication in deriving normal physiologic erythrocellular data; the development of the precursors of the erythrocyte (normoblasts); the morphologic analysis of the megaloblastic series of abnormal erythroid cells; the analysis of erythropoiesis in bone marrow; the relationship of the simian immunodeficiency virus and erythropoiesis; the erythrocyte life span; and the parasitic invasion of the red cell. A major segment of this monograph is an extensive table (Table 1) of erythrocellular data that is primary source cited. Whenever offered in the original publication, statistically significant differences between the sexes, experimental conditions, and so on, are indicated. Table 1 serves as a source for specific information and documentation for many of the statements offered in the text.

Acknowledgments

The authors are indebted to their mentors, Dr. R. Dorothy Sundberg, Dr. John Rebuck (CAG), and Professor Francesco Della Corte (AP), who have led them to an appreciation of hematology. A thank you is due to André Nault, head of the Veterinary Medical Library, and librarian Mary Lisa Berg, the University of Minnesota, for sharing their productive, stimulating environment. Dr. Richard Smith, friend and colleague, made research support available. Dr. Thaddeus M. Szczesny's technological acumen made it possible to include the illustrations in the text.

Authors

Chester A. Glomski is a professor of anatomy at the State University of New York at Buffalo (SUNYAB) Jacobs School of Medicine and Biomedical Sciences. Dr. Glomski earned his PhD from the University of Minnesota and his MD from the University of Mississippi School of Medicine. His research interests include experimental, morphologic, and comparative hematology and the effects of low-dose X-irradiation on hemopoiesis. Dr. Glomski has received a number of awards, including the Alan J. Gross Award for Excellence in Teaching (2011). He has been an invited lecturer and presenter at various national and international conferences. He has also authored and coauthored numerous abstracts, chapters, and books, including the following two books with Alessandra Pica (one of the coauthors of the present book): *The Avian Erythrocyte: Its Phylogenetic Odyssey* (Science Publishers/CRC Press, 2011) and *Erythrocytes of the Poikilotherms: A Phylogenetic Odyssey* (Foxwell & Davies, 2006).

Alessandra Pica is an associate professor of hematology in the Department of Biology at the University of Naples (Federico II, Italy) and a consulting hematologist for the Sea Turtle Rescue and Rehabilitation Program of Zoological Station in Naples A. Dohrn. She has taught a number of courses, some of which include human anatomy, embryology, and experimental morphology; hematology and hemopathology of sea turtles; and hemopathology of seawater vertebrae. She received her bachelor of science degree from the University of Naples. Pica is a member of the following organizations: the Unit of Naples Project PRIN 2003, the Scientific Committee of C.I.R.U.B. (Centro interdipartimentale per l'ultrastruttura biologica) of the University of Naples, the Animal Experimentation Ethics Committee of AORN Cardarelli of Naples; the Italian Society of Anatomy, the Italian Society of Histochemistry, and the Italian Society of Zoology. In addition to publishing numerous articles, Pica, with Chester Glomski (one of the coauthors of the present book), coauthored *The Avian Erythrocyte: Its Phylogenetic Odyssey* (Science Publishers/CRC Press, 2011) and *Erythrocytes of the Poikilotherms: A Phylogenetic Odyssey* (Foxwell & Davies, 2006).

Jessica F. Greene received a PhD in pharmacology and toxicology from the University of California at Davis, and undergraduate degrees from The Evergreen State College (Olympia, Washington). She has numerous publications in the fields of toxicology and risk assessment. After many years of consulting for Exponent, Inc., Dr. Greene is currently the CEO of Greene Consulting in Portland, Oregon.

CHAPTER 1

Pursuit of the Baseline, Steady-State Hemograms of *Macaca mulatta* and *Macaca fascicularis*

Scientifically correct hemograms of *Macaca mulatta* representing standard, unperturbed physiologic conditions would be presumed to be derived from analyses when the following conditions are in place: the monkeys are healthy, captive, laboratory-bred adult, nongravid subjects of a single sex, perhaps 4–15 years of age. Single housing is likely preferable for many experimental circumstances. The subjects should have been familiarized with the test laboratory setting for perhaps 3 months prior to study. The monkeys should be successfully trained for handling and to be cooperative for a stress-free phlebotomy. The tests should be performed on fasting subjects (~12–16 hr) in the absence of pretesting sedative medication. The hemograms would ideally be determined more than once on a given individual or longitudinally monitored to establish mean values. The analyses should employ contemporary electronic methods.

Two investigations that conform to these conditions are those of Hassimoto et al. (2004) and Rosenblum and Coulston (1981) (Table 2). The particular benefits of the former researchers' study were that 3.5-year-old subjects (n = 6 of each sex) were subjected to six consecutive monthly analyses. They were thus 4 years old by the end of the study. Cumulative mean values were determined from the reported data. The monkeys had been singly housed and acclimated to the colony for 6 months that included daily body touching, daily hand-to-hand feeding, and monkey chair restraint placement for 10 minutes two times per week. A suggestion that the monkeys were in a stable environment that was conducive to the generation of accurate data is the observation made by the workers that only fleeting changes were observed in the erythrocyte counts, hematocrits, and hemoglobin values across the 6-month investigation. Hence, this study accumulated a total of 36 hemograms from six male subjects and an equivalent database from a comparably analyzed female population. In the case of Rosenblum and Coulston, the monkeys were identified as sexually mature and had been obtained from multiple institutional sources and were trained to present an arm through a small opening in the cage for blood withdrawal. They were fully alert and untranquilized. The population also conformed to the conditions

Table 2   Erythrocyte Counts and Related Values for Laboratory *Macaca mulatta* under Idealized Conditions

	RBC	Hct	Hb	Diameter	MCV	MCH	MCHC
M. mulatta[42] ♂, adult, 6 subjects, 36 analyses	5.89	43.0	13.8	—	73	23	32
M. mulatta[16] ♂, adult, 17 subjects, 102 analyses	5.89	—	15.4	—	—	26	—
Weighted mean	5.89	—	15.0	—	—	25	—
M. mulatta[42] ♀, adult, 6 subjects, 36 analyses	5.58	42.3	13.9	—	76	25	33
M. mulatta[16] ♀, adult, 15 subjects, 100 analyses	5.69	—	14.9	—	—	26	—
Weighted mean	5.66	—	14.6	—	—	26	—

Sources: The superscript reference numbers refer to the identity of the investigators in Table 1.
Notes: RBC, mean erythrocyte count (millions/µL); Hct, hematocrit; Hb, grams hemoglobin/ deciliter; Diameter, mean diameter (µm); MCV, mean cellular volume (fL); MCH, mean cellular hemoglobin (pg); MCHC, mean cellular hemoglobin concentration (wt./vol%, g Hb/100 mL rbc).

described earlier. The adult population in this investigation consisted of 17 males and 15 females. An average of five samples was analyzed per subject. Four of the males and four of the females were studied on a longitudinal basis. They were analyzed every 6 months for 3 years. It was reported that the values derived from the longitudinal tests fell within the ranges obtained from the nonlongitudinal data. The data from males and females were maintained separately. The hematocrits determined in Rosenblum's and Coulston's investigation were derived by the centrifugation technique and not by electronic methods and consequently were not considered in the comparisons of these two investigations. Since the calculation of the MCV and MCHC involves the value for the hematocrit, these indices were also not included.

The data generated by these two "ideal" investigations are detailed in Table 2. The values were proportionally incorporated according to the number of subjects in order to derive a grand weighted mean. Each individual monkey, as noted, participated in several samplings (Table 1). It is seen that the mean erythrocyte counts exhibited a 100% concordance in the weighted average for the males of the two studies (5.89×10^6/µL). The agreement for females' mean red cell counts for the grand weighted mean of the two investigations (5.66×10^6/µL) was also very close, that is, 99%. The mean hemoglobin values of the two studies and their representation in the weighted grand mean were less tight but nevertheless in reasonable agreement. For the males the concordance of the two experiments in the derived mean grand mean (15.0 g/dL) was 92% and 97%, respectively while among the females the levels of concurrence were 95% and 98%. As would be anticipated, the grand weighted MCH, an erythrocyte index that is mathematically calculated from the red cell count and hemoglobin level of each sex, is in close agreement for the comparable groups.

From an overall observation of the studies mentioned above it can be accepted that the derived values reflect the generally accepted descriptions of the red cells of *Macaca mulatta*. The erythrocyte count is high (nearly 6 million/µL), and the red cells are smaller than those of man as implied by the fact that the hemoglobin content of erythrocytes of the studied healthy monkeys is less than that of man (i.e., man ~30 pg/red cell versus rhesus monkey 25–26 pg/red cell). The calculated average size of the red cell, that is, the MCV (derived in Table 2 solely from Hassimoto et al.), is correspondingly low in comparison with man (monkey ~75 fL as opposed to 90 fL for man). The axiom that the male of *Macaca mulatta* is likely to maintain a higher red cell count and higher hemoglobin level than the female counterpart is in agreement with the results seen here. Specific comments regarding the hematocrit are not made because Hassimoto employed electronic methods for this parameter while Rosenblum and Coulston did not. It is recognized that the investigations under discussion are not of a significantly large population and that individual variations typically obtained in a broad normal population are not included in this sampling. Thus, the values derived here cannot be applied to the overall *Macaca mulatta* species. It is interesting, however, that these two investigations that were conducted in two different institutions almost 25 years apart derived such similar results when the monkeys were studied under the same controlled circumstances.

The long-tailed macaque (cynomolgus monkey) *Macaca fascicularis* is a fellow constituent of the Genus (i.e., *Macaca*) to which the rhesus monkey belongs. It is often enlisted as an equivalent for the rhesus monkey in studies where a small laboratory-maintained nonhuman primate is the experimental subject of choice. This monkey has its origin in Southeast Asia. Most studies involving this monkey (synonymously termed the crab-eating monkey) seem to be concentrated in Far East countries such as Japan and China. This species has been described as the most widely used investigational nonhuman primate probably on the geographical basis of its inclusion in biological investigation both in the Western and Eastern world. It is the most arboreal of all nonhuman primates and is anatomically unusual because it has an extraordinarily long tail that is almost always longer than its height from head to rump and ranges between 40 and 65 cm (~1.5–2 ft). Its arms are described as relatively short. The collected hematologic data of this species are listed in Table 1.

A representative blood picture of *Macaca fascicularis* is the following (5-year-old monkeys that were housed in indoor pens in same sex social groups that were formed when the animals were 6 months of age. Blood samples were collected at a biannual analysis from overnight fasting monkeys; they were humanely phlebotomized and not administered sedation. The analyses were conducted electronically): Male profile—Rbc $5.90 \times 10^6/\mu L$, Hct 46.9%, Hb 13.3 g/dL (n = 44); female profile—Rbc $5.32 \times 10^6/\mu L$, Hct 42.9%, Hb 12.1 g/dL (n = 47) (Table 1, Xie et al. 2013). Tables 5 and 6 present the weighted mean erythrogramic values for age–sex grouped *Macaca fascicularis*. Phlebotomies conducted in the absence of ketamine are included in Table 5, while those performed under ketamine sedation are listed in Table 6.

The adult rhesus and cynomolgus monkeys are roughly similar in size but these two species are nevertheless incrementally different in magnitude. The rhesus monkey is larger (♂ 5.5–12.0 kg, ♀ 4.4–10.9 kg) while the long-tailed macaque has a weight of 4.7–8.3 kg in males and 2.5–5.7 kg in females (Bernacky et al. 2002). Similar ranges have been reported by Fortman et al. (2002). Sexual maturity is achieved in *Macaca mulatta* males at 36–48 months of age and in females at 31–42 months of age. In the case of *Macaca fascicularis* in males, it occurs at 36–48 months of age and in females at 36–41 months of age (Fortman et al. 2002). Differences are also frequently but not always identifiable in the mean erythrocyte counts of the two species. By and large the erythrocyte count tends to be lower in *Macaca mulatta*. This is best recognized in non-ketaminized males in the 2–3 to the 10-year-old range of both species (Tables 5 and 12). It is more difficult to verify a difference in the dimensions of the individual erythrocytes of the two monkeys (mean erythrocyte volume, MCV). It would be anticipated that the size of red cells would be smaller in the cynomolgus monkey (as suggested by the Charles River Laboratories[71] subjects; Table 1) but the summaries of the readily available data do not support this concept. This might be demonstrable in studies designed to specifically address this question. Conventional analyses have shown that smaller animals have a higher metabolic rate than larger species, because at least in part, they have a greater total body surface area relative to their body mass than maintained by larger species. This relationship results in the need to maintain a higher metabolic rate to sustain this body exposure. This in turn leads to the need to have a higher erythrocyte count to supply sufficient oxygen to maintain an enhanced metabolic rate. This status is viewed as one that fosters the development of smaller red cells because in aggregate their sum total of surface available for oxygen exchange is greater than that of the same mass of larger red cells. In accordance with this reasoning, it would be anticipated that the smaller of the two compared macaques (i.e., *Macaca fascicularis*) would likely maintain a larger number of red cells per unit volume of blood than the larger monkey and that the cynomolgus monkey's red cells would in addition be smaller than those of the rhesus monkey. These relationships are also discussed in the section devoted to the life span of the red cell.

CHAPTER 2

Stress and the Erythrogram

It is universally recognized that stress can have an impact on the erythroid (and also leukocytic) values of the circulating blood. This is part of the alarm reaction that was first recognized by Selye (1936) and Harlow and Selye (1937). Investigators dealing with nonhuman primates are typically knowledgeable about the occurrence of modification of the blood picture due to stress and are also interested in its quantitative and durational aspects. In many cases, the alarm reaction (synonym: stress reaction) is the result of relatively straightforward occurrences of experimental studies such as the use of restraint or a squeeze-back cage, placement in a primate-restraint chair, the change from a social existence to a solitary caged life, or vice versa, relocation from one laboratory to another, the initiation of experimental studies on an individual monkey that has been and continues to live within its own cage (e.g., phlebotomies, intubations, injections), or temporary transfer from the home cage to a platform for a procedure followed by a return to the home cage.

The experimental requirement for rhesus monkeys (n = 18, 6–9 months of age) to be in a quiescent, relaxed, unexcited mode while being maintained in a supine position on an animal board for 1–2 hours and undergoing multiple phlebotomies can be cited as an example requiring trained, cooperative subjects (Gregersen et al. 1959). Some candidates proved to be unsuitable.

Several decades of stress research have demonstrated that social separation and exposure to a novel environment are two of the most reliable ways of eliciting behavioral or physiological indicators of stress (Capitanio et al. 2006). It is also well known, as Selye (1936) suggested in his earliest study, that "habituation or inurement can occur" and consequently the alarm reaction is not induced if and when the subject becomes accustomed to the adverse stimulus. Thus, for example, a modification of an erythrocyte count may not repetitively occur following phlebotomy once the procedure is no longer viewed as a threatening event (Figures 1 and 2).

Acute stress, excitement, and exertion result in the release of erythrocytes from the spleen and a consequent increase in the number of circulating red cells. A change in the concentration of red cells in the plasma along with concomitant changes in the hematocrit and hemoglobin (typically recognized as hemoconcentration) may also occur secondarily to the effects of epinephrine or other factors as a part of the fight

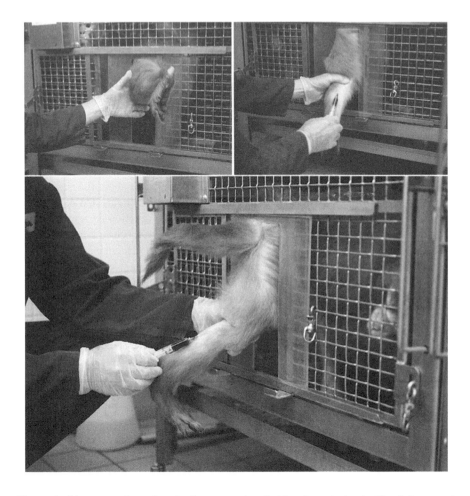

Figure 1 *Macaca mulatta*. A male rhesus monkey that has been trained with reinforcement techniques to actively cooperate during phlebotomy in its own home cage. Once trained, a monkey will cooperate with any person experienced in working with rhesus macaques. (From Reinhardt, V. and Reinhardt, A., *Environmental Enrichment for Caged Rhesus Macaques*, 2nd edn., Animal Welfare Institute, Washington, DC, 2001.)

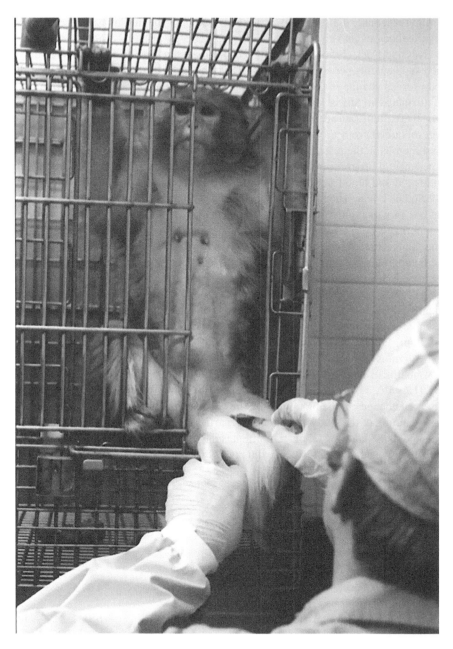

Figure 2 *Macaca mulatta*. A trained, compliant monkey voluntarily participating in a femoral phlebotomy. Noteworthy is the fact that under such conditions one unassisted caretaker can routinely obtain the blood sample. The most common research procedure, to which rhesus monkeys are submitted, is venisection from the saphenous or femoral vein (as illustrated). (From Reinhardt, V. and Reinhardt, A., *Environmental Enrichment for Caged Rhesus Macaques*, 2nd edn., Animal Welfare Institute, Washington, DC, 2001.)

or flight response. It also results in a redistribution of the marginating and extravascular leukocytes into circulation.

Ives and Dack (1956) conducted an experiment that involved the daily feeding of food supplements per stomach tube to 36 young *Macaca mulatta* and proposed that an observed change in the hemogram during the course of their investigation could be attributed to stress at the initiation of the study and its subsequent disappearance thereafter (Table 3). In this study, 18 male and 18 female ~3-year-old macaques were obtained from a conditioning farm (in South Carolina) and then acclimated for 4 months in the test colony laboratory prior to the start of the investigation. The experimental program involved daily placement in a special restraining box for the intubation and blood sampling (from the ear) that occurred at 3-month intervals for a period of 1 year. The monkeys became accustomed to being placed in the restraining box in which only the head was exposed and the body was otherwise relatively immobilized by foam rubber pads without straps or restraint from human hands. Once thus situated, they would patiently wait for the procedure. A comparison of the hemograms obtained at the start of the study and at 3 months later revealed a decrease in the erythrocyte count and hemoglobin concentration in the second sampling (Tables 1 and 3). Hematocrits were not determined. A decrease in the leukocyte count was also observed. In regard to the latter, the absolute neutrophil count was essentially unchanged in the two analyses while the lymphocyte count dropped from 10,126/μL for the initial analysis to 3,385/μL 3 months later. The mean values for erythrograms obtained for the 6–12-month duration were identical to those derived at 3 months (3 months: Rbc 3.96×10^6/μL and Hb 11.6 g/dL versus 3.98×10^6/μL and 11.7 g/dL, respectively). These results suggest the following interpretation. The observed changes in the blood picture are consistent with the occurrence of stress at the beginning of the study and at least its partial amelioration with the experience of repeated daily handling (about 4–5 minutes/day per monkey). It is seen that if indeed the 4-month acclimation residence did possibly decrease the level of stress perceived

Table 3   Erythrograms of Young Adult Rhesus Monkeys Obtained during and in Absence of Stress

Source	Rbc	Hct	Hb
Ives et al. (1956)[23], acute stress, handling, 18♂, 18♀	4.32	—	13.4
First sampling			
Second sampling, 3 months later, accustomed to handling, no stress	3.96	—	11.6
Lilly et al. (1999)[24], acute stress, capture, 45♀	4.63	35.8	10.7
First sampling, ketamine			
Second sampling, 1 month later, accustomed to sole caging, presumed decreased or no stress, ketamine	5.55	41.3	12.5
Lilly et al. (1999)[24], chronic stress of sole caging, 45♀	5.37	41.3	12.3
Five samplings in 6 months, ketamine			
Transfer to small group caging, improved cage, samples at third and seventh month after transfer, presumed no stress, ketamine	5.36	41.3	12.5

Sources: Superscripts refer to the identity of the investigators as cited in Table 1.
Note: Mean values, Rbc = $\times 10^6$/μL; Hct = %; Hb = g/dL.

by the monkey when it was placed in the holding box, intubated, and subjected to blood sampling, situational stress was nevertheless perceived by the subject during these latter procedures. This is in accord with the higher red cell count and hemoglobin level and a return to "normal levels" within 3 months. As a point of reference, during the time of the probable greatest stress (start of the feeding program and first blood sampling) the mean red cell count was 9% higher and the mean hemoglobin concentration was 16% higher than when stress was absent (or moderated) at the sampling 3 months later. Statistical analysis indicated an absence of differences in the data that could be attributable to sex of the subjects thereby helping make the evaluation of the study straightforward.

Phlebotomy at a site away from the "safe and secure" home cage can be viewed as a stressful event even when the test subjects have been habituated over several years to conventional blood drawing at a familiar remote locus (Reinhardt et al. 1991). Six adult male rhesus monkeys (9–13 years of age) were well familiarized with a procedure whereby each individual was removed from its home cage, placed in a transport box, transferred to a restraint apparatus located in another room, and phlebotomized. There was no necessity to force the monkeys to enter the transport box or the restraint apparatus. Further, it was not necessary to immobilize the monkeys in the restraint apparatus since they actively presented a leg for a saphenous venipuncture. After drawing of the blood, the given animal was returned to its cage and rewarded with favored food. The same monkeys were later trained to undergo cooperative phlebotomy in their home cage. They were trained to present a leg through an opening in the cage and submit to a saphenous venipuncture. (The training procedure of indoctrinating monkeys in voluntary cooperative stress-free phlebotomies is discussed elsewhere in the text.)

In the experiment with the aforementioned monkeys, blood samples were collected by the caretaker with whom the animals were familiar. Each participant was bled using the older conventional method (transfer to the restraint apparatus in another room, etc.) and the new procedure (in-home cage) on different days at 12:00 hr and again at 12:15 hr. The serum cortisol levels were determined on all samples (and used as an indicator of stress that the subject is undergoing). The levels of serum cortisol did not differ significantly between the first samples obtained at the remote site and the first samples obtained at home cage phlebotomy. However, cortisol concentrations were significantly higher ($P < 0.05$) during the second venipuncture (12:15 hr) in the restraint apparatus than during the second blood drawing in the home cage. In the case of home cage venisection, the second phlebotomy presented cortisol levels that were in the mean 13% higher than the first sample (not significantly different). Conversely, the second phlebotomy at the restraint apparatus offered a concentration of cortisol that was a mean 52% higher than the first sample at this site. This difference was statistically significant, $P < 0.025$. It is thus seen that the same cooperative monkeys that were familiar with phlebotomy away from their home cage nevertheless perceived this procedure as more stressful than in-home-cage blood drawing (as indicated by serum cortisol levels). A habituated blood sampling at a familiar but away from home locus was translated by the monkey as more threatening than the same process at the home site.

A roughly equivalent investigation with adult female rhesus monkeys corroborates these conclusions (Reinhardt et al. 1990). Ten adult females that were accustomed to phlebotomies performed at a site other than their home cage were observed to yield significantly enhanced levels of serum cortisol in the second of paired venisections performed 15 minutes apart. The increase was 50%, and it attained statistical significance, P < 0.001. These subjects, like their male counterparts discussed earlier, did not require force to cause them to enter the transfer box or the restraint apparatus. It was not necessary to immobilize them in the restrainer since they readily presented a leg for venipuncture. On the other hand, five other equivalent female monkeys that were trained for cooperative, voluntary phlebotomy in their home cage displayed cortisol values in the first phlebotomy of the same order as that obtained from the monkeys sampled in the "away from home" restraint apparatus but had a lesser increase of 18% in the level of the hormone in the second venipuncture. The increment did not attain a level of significance (P > 0.1). This investigation thus supported the premise that in-home-cage phlebotomies are less stressful for the monkey than venisections elsewhere. These results also verified that both male and female adults of the species maintain this characteristic.

Rhesus monkeys (nonpregnant adult females housed in individual squeeze-back cages) whose hematocrits were obtained while under ketamine anesthesia and also without the use of the anesthetic 2 weeks later displayed significantly higher hematocrits when tested without the use of the drug (Loomis et al. 1980, Table 1; also discussed elsewhere in the text). In another investigation of mature female *Macaca mulatta* monkeys, erythrograms obtained before and 15 minutes after ketamine injection yielded significantly higher Rbc, Hct, and Hb levels in the samples obtained prior to the administration of the medication (Bennett et al. 1992, Table 1; also discussed elsewhere in the text). In this instance, the degree to which the subjects were stressed was not apparent. Though restrained by two handlers during phlebotomy, these 16 adult female *Macaca mulatta* were described as conditioned to handling and appeared calm during the venisection procedure. Nonetheless, the pre-ketamine values, as noted, were higher. In another laboratory, ketamine-free hematologic evaluations of nine infant monkeys were conducted when they were 46, 50, and 58 weeks of age (2–4 analyses per individual) and compared with the data obtained from studies that utilized ketamine when the subjects were 42, 54, and 62 weeks old. The samplings that were obtained without the anesthetic consistently displayed significantly enhanced Rbc, Hct, and Hb levels (Fernie et al. 1994, Table 1; discussed elsewhere in the text). In these three foregoing studies, elevated values were derived from unmedicated subjects that can be assumed to have been more stressed than when they were analyzed under the influence of the anesthetic. Though these examples cannot be viewed as definitive demonstrations that a macaque under stress presents an enriched erythrocyte count, hematocrit, and hemoglobin level, the observed results do support this premise. It can be proposed that the use of ketamine is unlikely to render a subsequent erythrogram lower than physiologically normal and as a consequence the elevated non-ketamine values should be considered elevated at least in part due to a component of stress perceived by the subject.

An investigation that assessed the relationships of the erythrograms of 45 two-year-old female *Macaca mulatta* with acute and chronic stress was conducted by Lilly and her associates (1999, Table 1). This 2-year study consisted of three phases. Phase I involved the day of capture of the monkeys and their removal from a food-provisioned, natural, outdoor free range setting. It was considered a period of acute stress. Phase II consisted of single caging for 1 year and was considered an interval of chronic stress. Phase III involved caged social housing in small, stable same sex groups for an additional 7 months. It provided an improved, more natural environment with a variety of enrichment features such as a meshed side wall that opened to the outdoors, swinging perches, trapezes, and so on. Presumably, this milieu was the least stressful that could be feasible for caged captive monkeys. A total of eight blood samplings were conducted on each participant during the experiment. The first month of Phase II and the first 3 months of Phase III were devoid of testing allowing accommodation to the setting. The first phlebotomy was performed within 50 minutes after capture while the others were accomplished within 20–45 minutes after entry of the caretaker into the housing area. An average dose of 70 mg of ketamine was administered for Phases I and II venisections and was also presumably given prior to the two samplings of Phase III.

Blood analyses taken at the day of capture and 1 month later during single caging revealed a significant increase in all major erythroid parameters (Rbc, Hct, Hb) in the second assessment (Tables 1 and 3). The mean of all five samplings of Phase II did not show any significant difference from the mean initial analysis of this phase (Table 3). In addition, the average erythrogramic values of this period are identical to the mean of the two analyses conducted during the 7-month-long term of Phase III (Tables 1 and 3). It is thus seen that whatever effects the stress of capture imposed on the first hemogram of the investigation, the subsequent analyses were repeatedly consistent and stable, and yielded a bona fide "enriched" erythroid picture. It is therefore apparent that the erythrogramic values obtained during the period of acute stress reflected more than acute stress because the subsequent data presented higher values. The usual profile for stress is hemoconcentration and other factors leading to elevated Rbc counts, hemoglobin, and hematocrit levels followed by lower values when normal conditions return. It is also assumed that the use of ketamine with these monkeys modified the erythroid response seen in the acute stressed and nonstressed states. (The leukocyte counts obtained in this study corresponded with those of Ives and Dack [1956]. That is, the total leukocyte counts were increased in both investigations during the period of acute stress. Additionally, the B-lymphocyte count was significantly highest during Phase I and gradually diminished but remained significantly elevated for as long as 28 weeks of single caging.)

The experimental design of Phases II and III permitted the comparison of the blood pictures obtained under chronic stress (single caging with visual and auditory contact with a limited portion of the other subjects) and under stress-free ideal conditions (each monkey was housed in a social environment with four other females in a cage with a wall that opened to the outdoors and was equipped with swinging plastic barrels and other simian-friendly amenities). As noted, it was seen that the mean blood profiles for Phases II and III are identical (Tables 1 and 3). From an erythropoietic viewpoint, for at least this specific set of circumstances, the so-called chronic stress of single housing yields the same profile as that derived from the more natural housing of five females occupying more desirable caging.

The level of plasma prolactin is seen to be significantly higher in monkeys during the acute stress of capture and chronic stress of long-term single housing than during "ideal" social caging (Lilly et al. 1999). Thus, increased levels of prolactin may be an indicator of some forms of stress. Yet, it is seen that the mean erythrograms of chronic stress and "ideal" social cage life can be comparable (Phases II and III of this investigation). This would seem to indicate that certain levels of chronic stress perceived by the subject are not necessarily reflected in its erythroid blood picture. Capitanio et al. (2006) pointed out that different measures of outcome can be affected differently by particular procedures. The normalization of one particular class of responses (behavioral, hormonal, etc.) does not necessarily imply normalization of all classes of responses.

In the experiment of Hassimoto et al. (2004) (also discussed previously), the blood and electrocardiogram profile of 3.5-year-old captive-bred rhesus monkeys (n = 6 of each sex) were monitored initially and thereafter monthly for 6 months. The monkeys were acclimated, as noted, by daily body touching, daily hand-to-hand feeding, 10 minutes of monkey chair restraint twice weekly, and twice weekly forced oral administration of water with a latex catheter. The monthly blood analyses and electrocardiograms were conducted while in the restraint chair. The electrocardiograms were run on the same day as the phlebotomies. The heart rate was initially rapid but it diminished with acclimation. It took 3 months to obtain a significant decrease in the heart rate. As a consequence, it was interpreted that the monkeys revealed they were under stress during the analyses for at least the first 3 months of the study. It is noted, however, that the 6 monthly hemocellular analyses were consistent and stable with solitary occasional deviations that were significantly different from the initial preacclimation value. Hence, it seems that the erythrogram may remain at the normal unstressed level in the rhesus monkey even in the presence of some "stress" as indicated by tachycardia.

As would be expected, members of the Genus *Macaca* other than the rhesus monkey also display alterations of their blood picture when they are under stress. An example is seen in the case of *Macaca fascicularis* cynomolgus monkeys (5♂, 3–4 years old) who underwent the disruption of transportation from Japan to Korea (along with dehydration) and installation into new laboratory housing (Kim et al. 2005a). They were phlebotomized on arrival and weekly thereafter over a period of 35 days (Table 1). The subjects were sampled within 30 minutes of ketamine administration. The erythrocyte count and hemoglobin concentration were significantly higher at the initial sample in comparison with each of five subsequent samples. The first hematocrit was also significantly higher than that of the sampling 1 week later. All other samples were consistently lower than the first but the differences did not attain significance. The occurrence of stress was supported by a statistically significant elevated level of serum cortisol in the first sample versus all others and a concurrent significant decrease in the neutrophil to lymphocyte ratio during the same interval.

The foregoing studies, though modest in scope, appear to yield support for certain concepts. When a rhesus monkey perceives it is in an acute stressful circumstance, its physiological response yields an elevation of the erythrocyte count and hematocrit and hemoglobin concentration. This response is identifiable in monkeys that are in the first year of life as well as in older mature individuals. It would seem that the rhesus monkey requires at least 1 month (perhaps 3?) to accommodate to a given organizational or procedural milieu before it is accepted as nonthreatening and yields a "normal" blood picture. It is seen, as routinely accepted, that individual caging of a monkey even for extended periods is compatible with the derivation of stable erythrograms (as long as the event of a phlebotomy is not perceived as a threatening event). The investigation of Ives and Dack demonstrates that daily placement of a monkey into a restraining box and be subjected to a brief manipulation while therein can become an accustomed routine procedure and not be a source of stress.

Just as the modified internal milieu associated with stress may result in an alteration of a derived erythrogram, other physiologic or biochemical situations can similarly cause varied blood pictures. An example is seen in the course of the anemia that occurs in association with deficiency of pyridoxine (vitamin B_6) in the rhesus monkey (Poppen et al. 1952). When placed on a pyridoxine-deficient diet, the subject develops a progressive drop of the erythrocyte count and hemoglobin within 2 months. The anemic state progressively becomes worse throughout the first 25 weeks of deficiency of the vitamin and will continue until the animal expires. However, at about 35–45 weeks of experimentation, the subject demonstrates a transient, impermanent "improvement" in the red cell count and hemoglobin in which the mean values of these parameters begin to return toward normal. The anemic state, nevertheless, does not disappear and in the long term continues to persist onward to fatality. Thus the erythrogram derived at the improved point suggests a temporary upgrading of the red cell count and hemoglobin level (presumed hemoconcentration), but in reality the total number of erythrocytes and the hemoglobin content of the body are believed to be not increased but rather the data reflect changes in the internal milieu of the body perhaps related to changes in the liver. This modified environment yields an altered concentration of erythrocytes and Hb contained in a unit volume of the circulating blood. Hence, an erythrogram derived at this point does not deliver the same assessment that is obtained from an erythrogram from an unperturbed monkey. This interpretation is likely to be correct in view of the fact that a decrease in the two parameters occurs if a modest dose of pyridoxine is administered during the time of "transient improvement." That is, the blood picture becomes worse and returns to the prior less healthy state (i.e., a lower Rbc and Hb) and the disease progresses. Here, it is believed that the total body load of erythrocytes has also not diminished but the volume and perhaps content of plasma have been altered.

CHAPTER 3

Ketamine: Its Use in Obtaining Erythroid Values of Blood

The use of the rhesus monkey *Macaca mulatta* (and other nonhuman primates) in hematological investigations requires consideration of the milieu in which blood samples are obtained. It is accepted that the rhesus monkey is not a naturally compliant, receptive subject when phlebotomies are desired. The behavioral response of female *Macaca mulatta* to human observation and handling procedures, for example, have been assessed as primarily aggressive (while the responses of *Macaca fascicularis* were primarily fearful in nature and those of *Macaca radiata* indicated moderate fear) (Clarke 1986b). There is also the concern that the handling of excited, unbridled monkeys can lead to the injury of the handlers as well to the animal itself. It is also understood that there is a scientific desire to obtain biological samples that reflect the normal, unstressed physiologic state. As a consequence of this set of circumstances, the medication ketamine (administered as its hydrochloride salt) has become the most widely used agent to sedate monkeys during phlebotomy and concomitantly derive "acceptable" data. Ketamine, 2-(chlorophenyl)-2-(methylamino) cyclohexanone, is a phencyclidine derivative that produces a state of dissociative anesthesia by interrupting certain association pathways in the brain before producing somesthetic sensory blockade (Castro et al. 1981). It is administered intramuscularly; it is rapid acting and has a prompt recovery. In the typical dosage of 10 mg/kg of body weight, the drug has an anesthetic effect of about 20–30 minutes. The agent has been useful with primates in the wild because it can be administered by blowgun dart while the animals are free in their natural environment (Roney 1971).

Loomis and his coworkers (1980) were among the first to specifically attempt to identify the effects of ketamine on erythroid values in the rhesus monkey. They injected ketamine into the muscles of adult female monkeys when they were immobilized with the squeeze mechanism of their individual home cages. A limb was manipulated through the bars of the cage, and ketamine was administered. Simultaneously, a sample of blood was obtained from either the saphenous or cephalic vein and also 10 and 20 minutes later. Control samples were taken from the same animals in the same manner without giving the drug. Control and experimental samples were taken 2 weeks apart. Twenty minutes after injection, a decrease in the hematocrit was identified in the sedated monkeys (in comparison with the

15

control group, $P < 0.10$). The mean level at that time was 40.1% while the mean level for the control samples acquired when the monkeys were physically restrained but not medicated was 43.3%, a difference of 7.4% ($n = 20$, Table 1). Erythrocyte counts and hemoglobin concentrations were also determined but were not reported. (Additional alterations induced by this agent included a reduction of circulating lymphocytes and to a lesser extent the neutrophils and a diminished level of plasma proteins.) A subsequent evaluation of ketamine-associated modifications of the erythrogram of 16 individually caged, healthy adult, nonpregnant females was conducted by Bennett et al. (1992) (Table 1). Samples of blood were obtained prior to and approximately 15 minutes after intramuscular injection of the ketamine. Statistically significant reductions in the erythrocyte count, hematocrit, and hemoglobin concentration were obtained following administration of the drug ($P < 0.001$, Table 1). The decrement was 8% for each parameter. The decrement, according to Bennett and coworkers, was attributed to the redistribution of the erythrocytes from the circulating blood to the spleen and extravascular sites. (Additional statistically significant non-erythroid changes included decreases in circulating lymphocytes [−51%], serum glucose [−20%], total serum protein [−12%], and albumin [−11%].) The biochemical alterations, according to the latter investigators, suggested an influx of fluid into the vascular space. Another independent material observation was that nine infant rhesus monkeys (ages 42, 54, and 62 weeks) that were administered ketamine and subsequently subjected to femoral venisection presented significantly lower red cell counts, hemoglobin concentrations, and hematocrits than when they were phlebotomized at 46, 50, and 58 weeks but not given the sedative (Table 1, Fernie et al. 1994). (The levels of circulating lymphocytes were likewise diminished [$P < 0.001$].) The cohort was comprised of five males and four females. It is assumed that the sex of the test infants did not have any effect on the hemograms as these monkeys were too young to present sex-linked erythroid divergence. Fernie and coinvestigators have pointed out that the described erythroid pattern was consistently displayed by each of the nine infant participants.

An examination of the effects of phencyclidine (ketamine is a derivative of this drug) on the hemogram of six male and six female *Macaca mulatta* (weight 4.0–6.1 kg, hence presumed to be mature monkeys) was conducted by Hemm and Johnson (1978). The monkeys were acclimated to the laboratory environment, housed individually, and fasted 16 hr prior to treatment (either 1.0 mg/kg phencyclidine or a comparable volume of saline). The subjects were immobilized in a squeeze cage, removed, and held on a restraining podium for all sampling intervals. After a 0 hour initial femoral sampling, the monkeys were immediately administered the phencyclidine or saline. Successive blood samples were withdrawn at 1 hr and 24 hr later. This regimen was conducted during weeks one, three, and five. Analyses revealed a graded reduction in the values for the Rbc count, Hct, and Hb concentration over the 5-week study that was similar in the control and phencyclidine groups. This change was attributed to blood loss from repeated sampling. However, phencyclidine per se caused no change in the erythroid values. It is noted that the first samplings were made 1 hr postinjection of the drug rather than 15–20 minutes postinjection as cited in the studies that utilized ketamine. In addition to the difference in the timing of

Table 4 Comparison of Typical Ketamine-Associated and Non-Ketamine Mean Erythroid Profiles of Adult, Male, Laboratory-Housed Rhesus Monkeys

	Source	No. of Subjects	Rbc	Hct	Hb
Ketamine	Hom et al. (1999)[43]	53	5.32	38.9	12.6
No ketamine	Stanley et al. (1968)[5], Lewis (1977)[48], Matsumoto et al. (1980)[8]	52	5.88	46.8	13.8

Sources: The superscript reference numbers refer to the identity of the investigators as cited in Table 1.
Notes: Mean values, Rbc = $\times 10^6/\mu L$; Hct = %; Hb = g/dL. Non-ketamine values, weighted mean of the three investigators' subjects.

the postinjection sampling, it is also recognized that although phencyclidine and ketamine are related compounds, they are not identical and can thus have varying physiological effects on the rhesus monkey.

An implication of the reductive effects of ketamine on the erythrogram can be obtained in comparing the erythroid values derived by Hom et al. (1999) who administered this agent to their monkeys prior to phlebotomy with that of other investigators who analyzed equivalent subjects without the use of ketamine. In Hom et al.'s investigation, 53 male rhesus monkeys were variably examined under ketamine anesthesia during a period of 4 years. Each subject was analyzed 2–15 times. As indicated in Table 1 and Table 3, their mean values were as follows: Rbc 5.32 × $10^6/\mu L$, Hct 38.9%, and Hb 12.6 g/dL. On the other hand, the combined weighted mean erythroid values for the males from the studies of Stanley et al. (1968), Matsumoto et al. (1980), and Lewis (1977), which were selected as having comparable monkey populations (age and experimental conditions, e.g., laboratory housing and methodology) but not administered ketamine were Rbc 5.88 × $10^6/\mu L$, Hct 46.8%, and Hb 13.8 g/dL (Table 1 and Table 4). (*Note*: The hematocrit data include only the values derived from electronic methods. Hematocrits obtained by the microhematocrit centrifugation technique were not included in order to maintain uniformity of comparisons.) It is immediately seen that the data derived from the ketamine-injected monkeys have lower values for all major erythroid indicators than the animals not administered the anesthetic. This of course is consistent with the descriptions given previously for the effects associated with ketamine usage.

It has been further observed that ketamine may have an accumulative effect upon the derived erythrocytic profile (Lugo-Roman et al. 2010). Monitoring of the complete blood counts of eight adult rhesus macaques (laboratory maintained, individually caged, and under fasting conditions; 6♂ and 2♀) who were administered ketamine at the typical dosage of 10 mg/kg of body weight on three consecutive days prior to venisection on each of these days revealed a progressive reduction of each major erythrocellular value (Rbc, Hct, Hb). Thus, using the quanta obtained on day 1 as the initial post-ketamine level each subsequent injection of the drug was associated with a decrement of the level of every parameter (Figure 3 and Table 1).

The effects that ketamine (commercial titles: Ketanest, Ketaset, Ketalar) has on derived rhesus monkeys' erythroid values when it is used to assist in conducting efficient, atraumatic phlebotomies, judging from its frequent use, have been considered

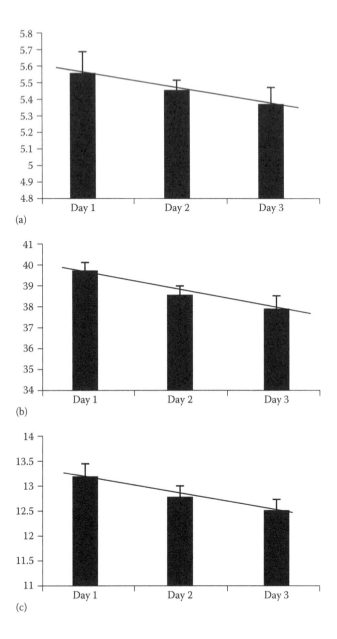

Figure 3 *Macaca mulatta*. Erythroid values following injection of ketamine into *Macaca mulatta* on three successive days. (a) Mean erythrocyte counts following injection of the drug (ordinate scale: 4.8–5.8 × 10⁶ Rbc/µL). (b) Mean hematocrits following injection of the drug (ordinate scale: 34%–41%). (c) The mean hemoglobin concentrations following injection of the drug (ordinate scale: 11–14 g/dL). Values ± S.D. It is seen that the injection resulted in a progressive reduction of the value of each erythrocyte-based parameter within the 3-day study period. (From Lugo-Roman, L.A. et al., *J. Med. Primatol.*, 39, 41, 2010.)

an acceptable cost (i.e., reduction in precision in reflecting the physiological state of the animal prior to the administration of the anesthetic). Further, when it is given to both the control and experimental subjects, its effects upon the end results of an investigation are often neutralized or rendered negligible. Ketamine has multiple applications in veterinary and human medicine. In addition to analgesia and anesthesia, it has also been used to treat bronchospasm and certain forms of depression in man. When the drug is used to facilitate the collection of blood samples from monkeys ketamine's effects on blood cell levels have been by and large proposed to be a reflection of its depression or reversal of the alarm response (Loomis et al. 1980; Bennett et al. 1992). Excitement and exertion, as noted previously, result in the release of erythrocytes from the spleen and a consequent increase in the number of red cells in circulation. It also results in a redistribution of the leukocytes. The observed decreases in the Rbc, Hct, and Hb concentration (as well as diminished lymphocyte counts) associated with ketamine administration are thus described as attributable to the reversal of the alarm response. It is also proposed that even monkeys that appear unstressed at phlebotomy nevertheless maintain a certain level of excitement at this event and its impact on the erythrocyte count is negated by ketamine.

The premise that ketamine interrupts or prevents the response to acute stress is in accord with the observed relationship of the drug with the levels of cortisol in the blood. Increased levels of cortisol in the circulation are a recognized accompaniment of acute stress. Puri et al. (1981) observed an increase in serum cortisol in adult male rhesus monkeys (n = 9) that were physically immobilized by an intracage device and serially phlebotomized four times at 20-minute intervals (hence subjected to stress). In another laboratory, a persistent significant elevation of serum cortisol was generated by 10 female adult nongravid rhesus monkeys that were submitted to 2–4-minute restraints (per squeeze mechanism of the cage) and venipunctures at 30-minute intervals for 2 hours (Fuller et al. 1984). An experiment was designed to compare the responses of 10 adult female rhesus monkeys to a 15-minute period of manual restraint while in the home cage with that obtained after a 5-minute confinement in a transfer box (Line et al. 1987). The study was controlled in a way that each test participant was tested twice in the home cage segment and then twice in the transfer box moiety. Pretest and posttest blood samples were obtained and thus reflected the response not only to the mode of restraint but also to the pretest venipuncture. There was a nonsignificant increase in mean plasma cortisol in response to venipuncture and restraint in home cage (P = 0.21). In contrast, the response to phlebotomy and transfer box confinement yielded a highly significant increase of cortisol (P = 0.003). The difference in results was theorized to be due to the subjects' (sometimes recognizable) assessment of the transfer box event as an undesirable experience as well as one with the potential for additional adverse (stressful) experiences.

Another study involving singly caged adult female rhesus monkeys demonstrated how the attendant circumstances of a phlebotomy can affect the consequent serum levels of serum cortisol. The cortisol values of 10 monkeys that had been routinely subjected to the conventional venipuncture technique of restraint on a table (Figure 4), 10 other monkeys that were accustomed to venipuncture in a restraint apparatus, and a cohort of 10 monkeys that had been trained for in-home-cage phlebotomy were

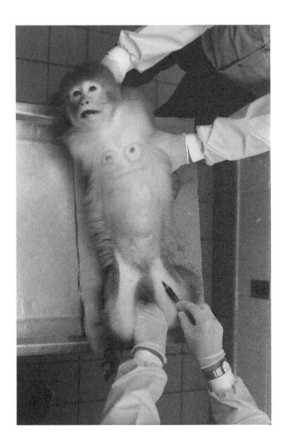

Figure 4 *Macaca mulatta*. Phlebotomy of a rhesus monkey while manually restrained on a table. The major undesirable feature involved in venipuncture of an uncooperative or untrained monkey is the physical restraint necessary to obtain the sample from the subject. The potential for an acute stress response to impact on a hemogram is recognized. Conventional venipuncture techniques can require two or three caretakers. Conversely, a single caretaker is often sufficient to conduct a phlebotomy on a trained monkey. (From Reinhardt, V. and Reinhardt, A., *Environmental Enrichment for Caged Rhesus Macaques*, 2nd edn., Animal Welfare Institute, Washington, DC, 2001.)

compared by Reinhardt (1992a, Figure 5). In the latter group, the subjects had been trained to allow the phlebotomist to pull a leg out of a partly opened cage and obtain a blood sample from the extended limb. Each animal from all groups was phlebotomized at the same time (13:15 hr) for the assessment of basal cortisol levels and again 15 minutes later (13:30 hr) for the assessment of cortisol response to the preceding venipuncture. The results revealed that all three groups of monkeys had statistically equal basal levels of serum cortisol while the levels for the second venipuncture differed in each cohort (Figure 5). The mean cortisol level for the animals bled on the table was significantly higher than that of the monkeys bled in the restraint apparatus ($P < 0.1$) as well as higher than that of bled in-home-cage monkeys ($P < 0.001$). The cortisol increase in the monkeys phlebotomized in the restraint apparatus though

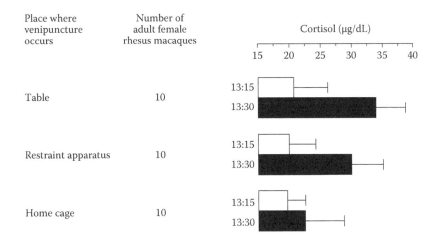

Figure 5 *Macaca mulatta.* Serum cortisol response to phlebotomy of three groups of adult female rhesus monkeys (n = 10 per cohort). Each group was routinely subjected to one of three different venipuncture procedures. The members of one group were restrained on a table during the procedure (as illustrated in Figure 4). The constituents of the second cohort underwent venisection in a restraint apparatus, and the third set of subjects were trained to voluntarily permit phlebotomy while remaining in their home cages. Conduction of venisection, while manually restrained on a table, was considered the most stressful technique while in-home-cage phlebotomy of a trained monkey was thought to be the least stressful. All the subjects were phlebotomized at the same time of the day and again 15 minutes later. The graph illustrates that the level of cortisol at the first phlebotomy was not significantly different in any group. The second phlebotomy revealed the effect of the first phlebotomy upon the level of cortisol as it was performed 15 minutes after the first bleeding. The mean cortisol concentration of animals, that bled on the table at the second phlebotomy, is seen to be significantly higher than that of subjects that bled the second time in the restraint apparatus ($P < .1$) or phlebotomized in the home cage ($P < .001$). All three categories of test animals show a cortisol response to blood collection. The magnitude of cortisol increase, however, was significant only in animals that underwent phlebotomy on the table or in the restraint apparatus. Thus, this experiment illustrates that the level of presumed stress associated with a mode of phlebotomy was accompanied by a proportional increase of serum cortisol. (From Reinhardt, V., Improved handling of experimental rhesus monkeys, in *The Inevitable Bond: Examining Scientist–Animal Interactions*, Chapter 10, Cambridge University Press, Cambridge, U.K., 1992, p. 175.)

higher than the basal level did not attain a concentration of cortisol as high as that of the monkeys bled on the table. On the other hand, the in-home-cage bled monkeys generated the least increment of the steroid. And it was not significantly different from the basal level. It can therefore be proposed that in this experiment the cortisol response was proportional to the amount of stress perceived by the monkeys. Phlebotomy on a table was assumed to be the most stressful procedure while the venipuncture of monkeys that were trained to submit to phlebotomy in their home cage was considered the least threatening. In addition, there is an implication that in-home-cage phlebotomy was nevertheless still perceived as slightly stressful because of the slight (statistically insignificant) increase in the mean concentration of cortisol

(Figure 5). Clarke (1986a) has noted the adrenocortical responses of female rhesus monkeys (as well as *Macaca radiata* and *Macaca fascicularis*) following handling, novelty, and restraint stress increased with the severity of stress exposure.

One unrestrained, active adult male rhesus monkey with a chronically indwelling jugular catheter and wearing a remotely radioactivated blood collection device supplied pairs of baseline stress-free blood samples (10 minutes apart) on five separate occasions. Immediately following a paired sampling, the monkey was captured and transferred by transport box to a cage where a sample of blood was collected by venipuncture of the saphenous vein (~18 minutes post-jugular samples) (Herndon et al. 1984). The activity associated with the capture and saphenous phlebotomy could be interpreted as a stressful event. Cortisol levels in the blood were consistently higher in the stress-associated samples. (The investigators indicated that a low dose of ketamine [1 mg/kg] was given to the monkey 4 hours prior to the experiment to fit the backpack containing the blood collection device on the monkey. This interval was selected to allow dissipation of the effects of the drug and handling of the subject.)

Wickings and Nieschlag (1980) observed that significantly lower levels of serum cortisol were generated following venipucture in accustomed-to-restraint-chair adult male *Macaca mulatta* (n = 4) when they were anesthetized with ketamine than when they were phlebotomized while conscious and had not been given the drug (Figure 6). Elvidge et al. (1976) have also documented that monkeys when anesthetized with ketamine present lower cortisol levels than when they are phlebotomized without prior ketamine. Puri et al. (1981) have noted that a single intravenous injection of ketamine into adult male rhesus monkeys results in a marginal increase of cortisol only at 3 hours postinjection. Winterborn et al. (2008) have described a lowering of the level of circulating cortisol following the administration of this anesthetic.

In contrast, the intramuscular injection of ketamine into *Macaca fascicularis* (the long-tailed or crab-eating macaque), a member of the same genus to which rhesus monkey belongs, apparently does not result in the same erythrocellular alterations that are achieved with the rhesus monkey. In one study of female *Macaca fascicularis*, the red cell, hematocrit, and hemoglobin levels were not significantly affected by ketamine (n = 6 experimental and 6 saline-injected control subjects) (Yoshida et al. 1986b). However, each of the levels of these major erythroid parameters decreased slightly following the administration of ketamine by 10 and 30 minutes postinjection. And at 60 minutes posttreatment, a decrement was typically still evident but to a lesser degree (Table 1). The magnitude of the modifications was definitely greater than those observed in the saline-injected controls whose changes would be attributable to the handling, venisections, saline injection, and so on. Further, a decrease in serum cortisol of the treated monkeys contrasted the elevated postsaline level of serum cortisol in the controls. The difference appeared to be the realm of statistical difference. This could be considered an inference that the ketamine injection served to ameliorate the stress reaction associated with the procedures. These responses to the medication suggest that the anesthetic does have some impact on the cynomolgus monkey's blood picture. In a more recent investigation by Kim et al. (2005b), with sedative and sedative-free analyses conducted 19 days apart, this species displayed significant mean ketamine-associated increases in the erythrocyte count (+6%, P < 0.001) and hemoglobin

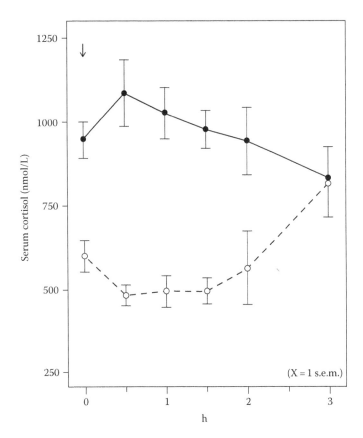

Figure 6   *Macaca mulatta.* A comparison of the level of serum cortisol in accustomed-to-restraint chair adult male rhesus monkeys (n = 4) that were phlebotomized following the administration of ketamine versus the level of this steroid in the same subjects when they underwent venipuncture when they were conscious and had not been given the medication. Closed (solid) circles indicate the cortisol levels in chair-restrained but unmedicated subjects. Open circles indicate the level of cortisol following the administration of ketamine while in the restraint chair. The concentration is recognizably lower following the administration of the anesthetic. The increase in the level of cortisol at 3 hours postinjection of the drug is postulated to be the result of a reduction in the depth of anesthesia. (From Wickings, E.J. and Nieschlag, E., *Acta Endocrinol.*, 93, 287, 1980.)

concentration (+7%, P < 0.001) in males (n = 19, Table 1). Females (n = 15) exhibited equivalent increments. The hematocrit declined in males (~1%) but the change did not attain significance while a decrement in females (−3%) was mathematically significant (P < 0.05, Table 1). The MCV was also reduced in both sexes (6%–8%, P < 0.001). While the males in the normal alert unanaesthetized state displayed minimally higher Rbc, Hct, and Hb values than their female counterparts, this differential relationship was also identifiable following exposure to ketamine. Both of the latter two investigations revealed a decrease in the white blood cell count after injection of ketamine.

Castro et al. (1981) have reported that ketamine given intramuscularly in 10 mg/kg and other dosages to chronically cannulated, restraining-chair-acclimated *Macaca fascicularis* produced no significant changes in their concentrations of plasma cortisol.

In an attempt to better discern the differences in the erythrograms that are derived from the blood samples of *Macaca fascicularis* when the subjects are sedated with ketamine as opposed to when blood is withdrawn in the absence of sedation, the readily available data were organized for appropriate comparison (Tables 5 and 6). The data were obtained from the reports cited in Table 1. The information was organized according to the subjects' age, sex, administration of, and absence of ketamine. The selection process also involved the identification of comparable subjects (e.g., health status, acclimation to a laboratory environment, nongravid status, etc.) and the use of modern contemporary electronic techniques. Hematocrits obtained by centrifugation and mean corpuscular volumes that utilized centrifuged hematocrits for their mathematical calculation were excluded from the data. The derived erythrocyte-based values (Rbc, Hct, Hb) are all mean weighted values for each specific age, sex, and sedation and nonsedation groupings (Tables 5 and 6).

In the ~1–2-year-old age range, the mean weighted erythrocyte count was definitely higher in both male and female ketamine-treated monkeys than in the nonmedicated counterparts. On the other hand, the Hct and Hb levels of non-ketaminized subjects of both sexes appeared to be substantially higher than that of ketamine-administered counterparts (Tables 5 and 6). It is noted that in these foregoing monkeys the males and females of a given experimental set (i.e., medication or absence thereof groupings) displayed virtually identical values between the males and females for a given parameter. Additional calculations that involved deleting the <1-year-old monkeys (Perretta et al. 1991) from the ketamine-treated <1–2-year group (thereby making the age range tighter and more homogeneous) did not modify the resultant averages for this cohort.

A comparison of ketamine-injected <1–2-year-old monkeys with the slightly more mature non-ketaminized 2.5–3.5-year-old monkeys (n = 35♂, 37♀) revealed the same relationships obtained in comparison of medicated and nonmedicated ~1–2-year-old animals. The erythrocyte counts of both untreated males and females were lower than that of ketaminized 1–2-year-old animals while hematocrits and hemoglobin concentrations of the older, nonmedicated monkeys were higher than observed in the medicated 1–2-year-old subjects (Tables 5 and 6).

An examination of the weighted mean erythrograms of the ketaminized and non-anesthetized 3–5-year-old long-tailed macaques reveals that the proportionality (bias) of the values observed in both sexes in the 1–2-year-old age range changes at this interim (Tables 5 and 6). The erythrocyte count is reduced (both sexes), and the Rbc count in the ketaminized and non-ketaminized subjects is equivalent. The single prominent relationship that is persistent is the higher hematocrit of monkeys not administered ketamine (hematocrit for ketaminized 1–2-year-old males and females = 42.7% and 42.8%, respectively [total n = 98], Hct for 1–2-year-old conscious males and females = 45.4% for each sex [total n = 653]; while the hematocrits for 3–5-year-old ketaminized males and females = 44.9% and 42.7%, respectively [total n = 85], and hematocrits for 3–5-year-old conscious males and females = 46.1% and 44.7%, respectively [total n = 320]).

Table 5   Age and Sex-Grouped Erythrocytic Profile of *Macaca fascicularis* Derived without Anesthesia

Age, Sex, Source	No. of Subjects	Rbc	Hct	Hb	MCV	MCH
Cesarean section, sex unstated, umbilical cord blood	6	4.98	44.1	12.7	89	26
Same subjects 5 hr later, femoral vein, Sugimoto et al. (1986)[20]	6	5.91	49.9	14.9	84	25
Birthday, full term, natural delivery, sex unstated	5	6.65	56.8	16.8	85	25
First week, daily 7×, 5 infants/per day, sex unstated, Sugimoto et al. (1986)[20]	35	6.55	54.1	15.9	83	24
1 mo old, sex unstated, Sugimoto et al. (1986)[20]	13	5.53	41.2	12.3	75	22
2–6 mo old, monthly 3–13 subjects, mean sex unstated, Sugimoto et al. (1986)[20]	7	6.87	45.5	11.7	66	17
7–11 mo old, monthly 3–11 subjects, mean sex unstated, Sugimoto et al. (1986)[20]	8	6.68	45.9	12.0	69	18
1 year old, males, Xie et al. (2013)[67]	162	5.78	44.9	13.0	78	22
1 year old, females, Xie et al. (2013)[67]	162	5.75	44.6	12.8	78	22
2 years old, males, Bonfanti et al. (2009)[60], Xie et al. (2013)[67]	136	6.01	46.0	13.2	77	22
2 years old, females, Bonfanti et al. (2009)[60], Xie et al. (2013)[67]	193	5.88	46.1	13.1	78	22
2.5–3.5 years old, males, Wang et al. (2012)[65]	35	6.34	49.7	14.5	79	23
2.5–3.5 years old, females, Wang et al. (2012)[65]	37	6.11	48.3	13.9	79	23
3–5 years old, males, Kim et al. (2005b)[12], Matsuzawa et al. (1994)[13], Zeng et al. (2011)[49], Xie et al. (2013)[67]	105	5.84	46.1	12.7	81	22
3–5 years old, females, Kim et al. (2005)[12], Matsuzawa et al. (1994)[13], Zeng et al. (2011)[49], Yoshida et al. (1986b)[66], Xie et al. (2013)[67]	215	5.65	44.7	12.5	80	22
5–6 years old, males, Matsumoto et al. (1980)[8], Xie et al. (2013)[67]	52	6.05	46.3	13.1	77	22
5–6 years old, females, Matsumoto et al. (1980)[8], Xie et al. (2013)[67]	55	5.52	42.7	12.0	78	22

Sources: The superscript reference numbers refer to the identity of the investigators as cited in Table 1.

Note: Rbc = ×10⁶/μL; Hct = %; Hb = g/dL; MCV = fL; MCH = pg. The values for each age–sex grouping are weighted mean values. Hematocrits derived by centrifugation were not included in the calculations.

Table 6   Age and Sex-Grouped Erythrocytic Profile of *Macaca fascicularis* Derived While Using Anesthesia (Ketamine)

Age, Sex, Source	No. of Subjects	Rbc	Hct	Hb	MCV	MCH
Less than 1–2-year-old males	49	6.51	42.7	12.2	66	19
Less than 1–2 year-old females, Sugimoto et al. (1986)[21], Perretta et al. (1991)[28]	49	6.49	42.8	12.2	66	19
1–10 years old, males and females, no significant differences identified between sexes. All ages had an equal distribution except the 2–3-year-old subjects that comprised 60% of the total population. Bourges-Abella et al. (2014)[62]	272	6.13	41.7	12.2	68	20
3–5 years old, males, Kim et al. (2005b)[12], Sugimoto et al. (1986)[21], Kim et al. (2005a)[53]	44	5.76	44.9	12.7	78	22
3–5 years old, females, Kim et al. (2005b)[12], Sugimoto et al. (1986)[21], Yoshida et al. (1986b)[66]	41	5.69	42.7	12.0	75	21
5–10 years old, males, Sugimoto et al. (1986)[21], Perretta et al. (1991)[28]	35	6.02	46.9	13.6	77	23
5–10 years old, females, Yoshida et al. (1989)[9], Sugimoto et al. (1986)[21], Perretta et al. (1991)[28]	353	5.99	41.1	11.5	68	19
8–18 years old, males, Sugimoto et al. (1986)[21], Verlangieri et al. (1985)[64]	41	5.86	38.5	12.3	66	21
8–18 years old, females, Sugimoto et al. (1986)[21]	17	5.72	44.1	13.0	77	23

Sources: The superscript reference numbers refer to the identity of the investigators as cited in Table 1.

Note: Rbc = ×10⁶/µL; Hct = %; Hb = g/dL; MCV = fL; MCH = pg. The values for each age–sex grouping are weighted mean values. Hematocrits derived by centrifugation were not included in the calculations.

Further examination of untreated 5–6-year-old male monkeys (n = 52) with ketaminized males 5–10-years-old (n = 35) revealed that these two groups offered equivalent values for the major erythroid parameters. The Rbc count for both groups was minimally increased but most likely not to a statistically significant degree. On the other hand, a higher hematocrit in the non-ketaminized 5–6-year-old females in comparison with medicated 5–10-year-old females was identifiable (but not statistically verified) (Tables 5 and 6).

In view of the cited aggregate of age and sex-indexed, weighted mean data concerning the erythrogram of the cynomolgus monkey *Macaca fascicularis* and the potential impact of the presence or absence of the anesthetic ketamine when obtaining this information, certain affirmations can be made. The most numerous differences in the erythroid values that are observed depending on whether or not ketamine is used to sedate this species of monkey are seen in young monkeys 1–2 years of age. On the basis of the aforementioned data, the treated monkeys of this age group displayed a higher red cell count while non-ketaminized subjects presented higher hematocrits and hemoglobin levels. These conclusions have yet to be statistically verified. It is also possible that the greater incidence of disparities between the erythrograms of the very young ketaminized and non-ketaminized monkeys as opposed to the less frequent presence of differences between older medicated and nonmedicated subjects could be due, at least in part, to differences in geographical origin and associated genetic makeup of the young monkeys. Once the monkeys have advanced to the juvenile level, that is, 3 years of age onward, the single individual modification of the erythrogram that is associated with the use of ketamine is a usually lower hematocrit (in comparison with test subjects of the same sex and age). As discussed previously, one might theorize that the higher hematocrit in the non-ketaminized specimens could be due to a stress reaction due to handling and phlebotomy which is otherwise modified by the administration of ketamine. Thus although the hemic picture of the rhesus monkey is more sensitive to the presence of ketamine than is the blood profile of the long-tailed macaque, the latter nonhuman primate nevertheless seems to have a minor version of the same reaction. Also as noted earlier, the occurrence of a physiologic reaction to ketamine in this so-called crab-eating monkey is identifiable in the lower level of cortisol in treated phlebotomized monkeys in contrast with the higher level obtained in phlebotomized but non-ketaminized subjects (Yoshida et al. 1986b). Ketamine does not alter sex-based differences in the erythroid blood picture that are concomitant with age and maturity.

CHAPTER 4

Training Rhesus Monkeys to Voluntarily Cooperate in Obtaining Phlebotomies

Rhesus monkeys are used in biomedical studies because of their close phyloge-netic relationship, physiological and immunological similarities to humans. They have been the most commonly used species of monkeys in biomedical research for decades (Capitanio et al. 2006). Consequently, more is probably known about the biology of this species than any other nonhuman primate. The results of investiga-tions with them, however, can be modified due to varying psychological and physi-cal stresses that are imposed upon these subjects in association with the studies. Immobilization of the monkey to obtain blood samples for hematologic assessments is a particular concern (Coleman et al. 2008). Erythrogramic values are subject to aberration during stress (the alarm reaction), and immediately reflect the current anxiety that is perceived by the subject. In many if not most circumstances, the actions and appearance of the monkey under scrutiny do not reveal the degree of stress that is present. Thus, the investigator is often unaware whether the derived data indicate an unstressed normal physiologic status or an altered condition. Some hor-mones normally present in the circulation have been identified as sensitive to stress (e.g., cortisol, testosterone, growth hormone, and prolactin).

Venipuncture is probably the most common handling procedure to which research macaques are subjected (according to a 10-year 1981–1990 survey of three princi-pal journals that published 397 original research articles on macaques) (Reinhardt 1991b; Reinhardt and Reinhardt 2001). The usual sites are the saphenous and femo-ral veins (Figures 1 and 2). The traditional method of conducting phlebotomies on monkeys has been to forcibly restrain the subject and then to obtain the blood sample (Figure 4). The stress involved in venipuncture has been said to lie primarily in the physical restraint necessary to obtain the sample (National Research Council 1998). It is accepted that this approach yields erythroid values that differ from those of the unperturbed animal. This imprecision has led to numerous studies to train monkeys to remain calm and unthreatened during phlebotomy as well as to actively cooperate in this process. Reinhardt (2003) has cited 10 different research facilities where macaques have been trained to voluntarily present a leg for blood collection (Figures 1, 2, and 7).

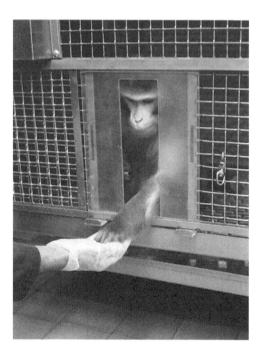

Figure 7 *Macaca mulatta.* As a part of the protocol to train rhesus monkeys to voluntarily
submit to phlebotomies, the trainers recommend that the subject should always
be rewarded with a favored food at the completion of the venisection. (From
Reinhardt, V. and Reinhardt, A., *Environmental Enrichment for Caged Rhesus
Macaques*, 2nd edn., Animal Welfare Institute, Washington, DC, 2001.)

The study of Ives and Dack (1956) is probably the most cited example of the
reduction of the erythrocyte count and concentration of hemoglobin in the blood as a
result of "training" the subjects. That is, the presumed normalization of the erythroid
profile as a result of the reduction of the level of anxiety perceived by the monkey.
This effect was attained by repeated handling and routine placement in a restraining
box. The scientific literature is replete with reports of investigations demonstrating
that *Macaca mulatta* that are trained to be phlebotomized and to cooperate with
the caretaker in this process generate lower levels of serum of cortisol following the
procedure than untrained counterparts undergoing equivalent blood sampling. The
monitoring of the levels of cortisol following a given manipulation such as venisection has become a generic assessment of the stress associated with the event. The
magnitude of the increment of the level of cortisol, within a given set of circumstances, has become an indicator of the relative amount of stress perceived by the
test monkeys (Clarke 1986a; Reinhardt 1992a). The work of Elvidge and coinvestigators (1976) offers an excellent example of how training of rhesus monkeys (adult
nonpregnant females) for phlebotomy yields significantly lower post-phlebotomy
levels of plasma cortisol than obtained from equivalent but untrained cohorts
(n = 6 per group). In addition, relative to the effectiveness of training, the trained
monkeys consistently had significantly lower levels of cortisol post-phlebotomy than

untrained monkeys given ketamine prior to phlebotomy. A similar response has been more recently reported by Bentson et al. (2003) for saline-injected and subsequently phlebotomized rhesus monkeys ($\male$, 10–13 years old). That is, untrained monkeys generated higher levels of cortisol than the trained counterparts. Further, untrained monkeys when injected with ketamine still offered greater concentrations of cortisol than trained monkeys not given the anesthetic but injected with saline. Lee et al. (2013) observed a statistically significant decrease in serum cortisol (as well as a decrease in the neutrophil to lymphocyte ratio) in female *Macaca mulatta* (n = 10, 3–4 years old) as they underwent a 5 times/week training for 1 month for accommodation to a monkey chair for experimental procedures.

There are several facts that are integral to the training of monkeys for nonstressful, voluntary, cooperative phlebotomy. The monkey is an intelligent, resourceful animal, and hence, it is a good candidate for training to offer a desired behavior. And under unfavorable circumstances, it is capable of behavior that is harmful to the caretaker and to itself. Positive reinforcement techniques (such as rewarding the monkey for a desired response with a favored food, e.g., raisins) are considered more effective in attaining intended end points than is negative reinforcement training in which the subject performs the desired action to avoid or escape from a negative stimulus (Coleman et al. 2008). Reinhardt, a scientist who has authored a score of publications regarding the benefits of training of monkeys to be cooperative subjects for phlebotomies, has indicated the importance of the trainee monkey to recognize and trust the trainer (Table 7). Winterborn and coworkers (2008) similarly state "the first step of shaping the desired behavior involves establishing a positive relationship between the nonhuman primate and the trainer." Of further interest regarding the training of a monkey is that once an individual monkey is trained to voluntarily cooperate in phlebotomy, qualified caretakers other than the trainer can obtain blood samples from the "educated" cooperative monkey as effectively as the trainer.

The physical activities that involve the handling of monkeys to obtain blood samples from them often involve two circumstances that trigger anxiety in the subjects. They are removal of the monkey from the home environment and the subsequent involuntary physical restraint. These problems are remediated first by training the monkey to be phlebotomized while it is in its home cage. This gives the monkey the opportunity to realize that it is not being removed from its secure familiar home and also affords it the freedom to independently move, if it chooses to do so. The second problem of involuntary restraint is eliminated by the training of the monkey to voluntarily offer a limb for the drawing of blood. Since this "learned" act is voluntary, it is not associated with a sense of fear. It has been shown that a trained monkey will cooperate in timed sequential venisections without a problem. Other than obtaining blood samples that reflect the current and unstressed condition of the subject, a trained monkey offers the significant advantage of being able to be phlebotomized by a sole phlebotomist (caretaker) who does not require the assistance of one or more individuals (Figures 1 and 2). A trained cooperative monkey can undergo blood sampling in one or two minutes.

The results of training rhesus monkeys ($\male$, 7–10 years old, n = 8) and chimpanzees *Pan troglodytes* ($\male$3, $\female$1, 16–43 years old) to be unstressed, cooperative participants

Table 7   Protocol for Training Rhesus Monkeys to Voluntarily Cooperate in Obtaining Phlebotomies

Step 1: The trainer should first establish recognition and a friendly relationship with the trainee *Macaca mulatta*. This may be accomplished by "visiting" the monkey to establish visual recognition of the trainer and offering a favored food such as raisins. The monkey should come to the front of the cage rather than retreat to the back when the trainer enters the room. It is recommended that this step should be accomplished prior to entering the training procedure.

Step 2: With the help of the squeeze-back mechanism of the monkey's home cage, the monkey is confined to the front quarter of the cage. In this position, the freedom of movement is considerably restricted, but the subject does have sufficient leeway to turn around when it assumes an upright stance. The monkey should be reassuringly talked to, gently scratched through the mesh of the cage, and offered some raisins. After a minute or two, the squeeze back is pushed back and raisins are offered again. After a few sessions, the animals learn that they are expected to stand during training. This exercise is repeated on different days until the subject is relaxed and accepts the food reward.

Step 3: The subject is restricted again and enticed with raisins and/or gently prodded to face the left or right side of the cage. The trainee monkey's leg is touched and groomed through the opening of the door of the cage. After a minute or two, the squeeze back is pushed back and raisins are offered. This sequence of events is repeated on different days until the animal stops retracting the leg and accepts the food reward.

Step 4: The restricted subject's leg is gently and firmly pulled through the opening of the door of the cage and firmly held for about 1 minute. The squeeze back is pushed back and the monkey is rewarded with raisins. The goal of this step is achieved when the animal shows no sign of resistance such as trying to retract the leg or turn around.

Step 5: The squeeze back is pulled only so far as to prompt the trainee to come forward. It is proposed that at this point the trainee monkey feels that it is in full control of the situation and has enough room to freely turn around and avoid being touched. The trainer encouragingly asks the subject to present a leg behind or through the opening of the door. An animal that refuses to cooperate is not punished in any manner but simply does not receive a food reward. This exercise is repeated on different days until the monkey actively presents a leg and shows no resistance during blood collection from the femoral or saphenous vein. Once this goal is achieved, the trainee is praised and rewarded with raisins (Figure 4.7). Both male and female monkeys are effectively trained with this procedure.

Although traditional manual restraint blood sampling procedures usually require at least two handlers, one to help restrain the subject and one to conduct the phlebotomy, only one person is required to perform a venipuncture on a monkey trained in the aforementioned protocol (Figures 2.1 and 2.2). Once a monkey has been trained, it will cooperate not only with the trainer but with other knowledgeable personnel. Only 1 or 2 minutes are required to withdraw a blood sample from a trained monkey.

Sources: Reinhardt, V., *Anim. Technol.*, 42, 11, 1991a; Reinhardt, V., *Lab Anim.*, 1, 32, 1996; Reinhardt, V., *J. Appl. Anim. Welfare Sci.*, 6, 189, 2003; Reinhardt, V. and Reinhardt, A., *Environmental Enrichment for Caged Rhesus Macaques*, 2nd edn., Animal Welfare Institute, Washington, DC, 2001, 77pp.

for venisection were compared (Coleman et al. 2008). Positive reinforcement techniques were used. The subjects were trained to place an arm in a "phlebotomy sleeve" and remain stationary for venipuncture (Figure 8). The protocol used by the investigators paired a primary reinforcer (a desired food) with a neutral stimulus (a click) to establish this latter stimulus as a secondary enforcer. The subjects were trained to perform various tasks by reinforcing successive movement toward the desired behavior. Each time a trainee performed an action successively closer to the target behavior, the trainer reinforced it by "clicking" and then rewarding with a food treat. The subjects were eventually trained to put their arm in the "phlebotomy sleeve" and hold a

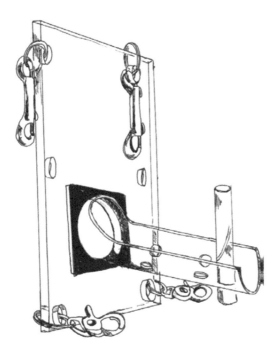

Figure 8 A phlebotomy sleeve used for training adult rhesus monkeys to cooperate in obtaining blood from a brachial vein. The apparatus was designed to cover the home-cage door. The armrest and the peg that the monkey grasps are apparent. (From Coleman, K. et al., *J. Am. Assoc. Lab. Anim. Sci.*, 47, 37, 2008. Illustration credit: J. Gregory Johnson.)

peg located at the end of the sleeve (to maintain a stable, firm nonmoving limb). To encourage the subjects to continuously hold the peg, the trainers provided constant access to a juice bottle with a favored drink (Figure 9). The training sessions for the rhesus monkeys were approximately 5–10 minutes in length (mean 5.1 minutes) 2–3 times/week. The training sessions for chimpanzees were 5–10 minutes (mean 7.3 minutes) 1–4 times/week. The length of the sessions was dictated by the attitude of the subject. Sessions were terminated if the subject lost focus on the trainer or became agitated. A monkey was considered trained when it remained stationary and allowed venipuncture in four of five sessions. All monkeys and chimpanzees were successfully trained to place an arm in the sleeve and remain stationary for venipuncture. However, repeated blood samples were only obtained from six of eight monkeys because one monkey had a clotting abnormality requiring extra attention during blood collection and so its training was discontinued while another monkey permitted the phlebotomy needle to be inserted but would pull its arm away from the trainer during the actual blood withdrawal. The total time invested in training the individual subjects varied across the populations. An average of 258 minutes (50 training sessions, over a period of ~7 months) was required to train each monkey. In comparison, a chimpanzee required an average of 219 minutes (31 sessions over a period of

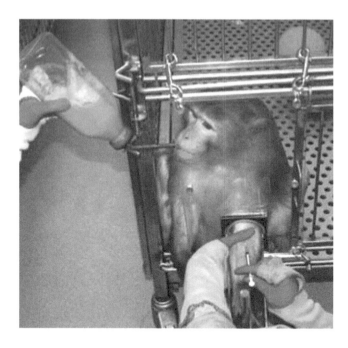

Figure 9   *Macaca mulatta.* An adult male rhesus monkey with an arm in the phlebotomy sleeve. While one trainer operates the syringe to obtain the blood sample from a brachial vein, another trainer provides the secondary reinforcement (clicking) and the juice bottle. The juice bottle is removed if the monkey does not perform the appropriate behavior. (From Coleman, K. et al., *J. Am. Assoc. Lab. Anim. Sci.,* 47, 37, 2008.)

~7 months). Although the total involvement of time required for training the individuals of both species was statistically similar, significantly more sessions were needed to train the monkeys than the chimpanzees. The training sessions were thus shorter for the monkeys than for the chimpanzees. As indicated previously, the length of the sessions was largely dictated by the subject's attitude. It would thus appear from this study that the rhesus monkey has a disposition that accommodates brief but not long training sessions. This study also demonstrated that the training of a rhesus monkey for a cooperative phlebotomy is roughly equivalent in effort to the training of a chimpanzee. The chimpanzee is acknowledged to be very trainable. The parallel study of chimpanzees and rhesus monkeys helps validate that the rhesus monkey can be trained to be relaxed and cooperative during phlebotomies.

This experiment also elicited the fact that the training for venipuncture using a phlebotomy sleeve may not be the routine, ideal approach for obtaining blood samples from macaques. The rhesus monkey has relatively small brachial veins consequently sometimes making collection of blood tenuous. Coleman and her coworkers (2008) reported that the phlebotomist was not always able to draw blood (locate and penetrate the vein) even if the monkey remained stationary during the entire procedure. Their animals were considered reliably trained when they remained stationary and allowed entry into the arm including redirections of the needle for four of five sessions. They proposed that the saphenous or femoral veins would be easier to work with in this species.

The training program of Coleman et al. (2008) to obtain blood samples from a brachial vein utilizing a phlebotomy sleeve that was appended to the monkey's home cage entailed a long duration (~7 mo) along with a considerable total investment of training minutes (~258 min/per monkey). Other protocols have been designed to offer in-home-cage phlebotomy with acquisition of the blood from the larger, more accessible saphenous, and femoral veins (Figures 1 and 2). These programs were also based on positive reinforcement techniques as advocated by Reinhardt who has been a long-standing proponent for this type of training for nonhuman primates (cf. reference list of Reinhardt and Reinardt 2001).

A relatively early study of the training eight female monkeys (5–16 years old) to cooperate in phlebotomies resulted in the completion of the training program in 24 work days and required a total of 250 minutes, that is, 31 minutes per animal (Vertein and Reinhardt 1989). On day 24 of training, blood collection was successively accomplished in all eight animals within a total of 17 minutes. Three monkeys actively presented a leg to the caretaker on this occasion. The other five monkeys were less cooperative but showed no signs of fear or resistance to having a leg pulled through the cage door and their saphenous vein punctured. All subjects accepted a food reward from the caretaker's hand after the venipuncture. A subsequent investigation involved the training of five singly housed adult male (mean 10-year-old) *Macaca mulatta* and utilized the experience derived from the prior investigation with female monkeys (Reinhardt 1991a). All animals had lived in their cages for more than 1 year. They were familiar with the trainer and took food from his hands. The monkeys were housed in squeeze-back cages, and this device was used in the training program to give the trainer access to the monkey as desired as well as to afford the monkey freedom of movement when it was part of the protocol. The trainer would also reach into the cage through an opening and gently touch and stroke the back of the trainee, grasp one of the animal's legs, and gently pull the leg out of the cage. It is seen that this was veritable training as initially the monkeys resisted being touched by repeatedly changing their positions or trying to reach back and slap, or even in one case try to jerk back the head and attempt to bite the trainer. Ultimately, each monkey proved to be trainable and cooperated during in-home-cage venipuncture not only with the caretaker but also other attending caretakers. One to two minutes were required to draw a blood sample. The average total accumulated time required to train an individual monkey to stop resisting the trainer's actions and allow the trainer to pull the leg out of the cage and take a blood sample was 32 minutes (range 9–65 min). It required a mean 44 minutes (range 21–69 min) for a given trainee to reach the end point of actively cooperating with the phlebotomy. The training sessions were ~3 minutes each. Training sessions were distributed over a period of 2–16 days. Five pairs of pair-housed monkeys of similar age that were also trained in this experiment yielded results that were comparable to those obtained with singly caged counterparts. A yet another example of training individually caged adult rhesus monkeys (n = 6♀, 8–12 years old) for in-home-cage voluntary cooperative phlebotomy was reported by Reinhardt (2003). The subjects were accustomed to being immobilized on a table for venipuncture. The monkeys were entered into the training program for cooperative voluntary in-home-cage phlebotomy and were successfully trained. Assessing the experience of this and other experiments that Reinhardt had conducted, the latter summarized that it takes a cumulative total of about 1 hour to train either a male or a female rhesus monkey for voluntary in-home-cage venisection. The effect of training on the level of serum cortisol was also determined in these subjects. Each monkey was phlebotomized twice on two separate occasions, initially at 13:15 hr and then at 13:30 hr under forced restraint on one day and then again on another day at-home-cage voluntary blood sampling. The mean cortisol concentrations of the first samples obtained under both conditions were statistically equal. Cortisol concentrations of the second samples, however, were significantly higher under restraint conditions than under the home cage process. In addition, the magnitude of the enhanced endocrine response to venipuncture was significant when the subjects were restrained (+68%, P < 0.001) but was not significant when they cooperated (+14%, P > 0.1). The premise that monkeys that are accustomed to conventional physical restraint phlebotomy conducted away from their home cage still perceive stress even though they have undergone the identical process many times appears to be affirmed. The occurrence of this anxiety can be verified by observing an increase in serum cortisol after an antecedent blood sampling. Table 7 presents a synopsis of the protocol for training voluntary cooperative in-home-cage phlebotomy.

An interesting investigation explored the feasibility of training very young juvenile *Macaca mulatta* for cooperative in-home-cage phlebotomy before they attained the point in their maturation when they would display aggressive behavior when they were physically handled (Reinhardt 1992b). The candidates were 14 females 13–18 months of age. They were housed in squeeze-back cages in pairs of the same age. The animals were screened for the project by restricting them to the front quarter of the cage with the squeeze-back mechanism and attempting to catch each subject by hand. Of the 14 juvenile macaques, 8 attempted to bite the investigator during the screening process. These animals ranged in age between 16 and 18 months. The remaining six juveniles were 13–14 months old. None tried to bite. These passive juveniles were selected for training for in-home-cage voluntary, cooperative venisection. The protocol was the same as the one utilized with the adult male rhesus monkeys described earlier (Reinhardt 1991a). The training sessions lasted 30 seconds to 5 minutes. It is noteworthy that the training was successful in only two cases. The two were cage mates. Their learning times were essentially identical. One subject voluntarily extended a leg through the cage opening after 47 minutes of interaction (38 sessions). The other presented a leg at the opening of the cage after an equivalent time. The long-term indelibility of their training was exhibited by the fact that both monkeys underwent successful triweekly testings during the subsequent 11-month follow-up period. Neither tried to bite or escape on these occasions. In the case of the other trainees, the training was discontinued with one pair after they stopped resisting having a leg pulled out of the cage (33 minutes, 24 sessions and 43 minutes, 34 sessions, respectively) but invariably refused to cooperate further. After a total of 120 minutes distributed over 75 sessions, neither of the monkeys was willing to present a leg for venipuncture. Their training was terminated at this point. The two monkeys of the third pair were afforded each with an investment of 60 minutes of training during 42 sessions. They did not overcome their fear (indicated by temporary diarrhea, grinning) and vigorously resisted having a leg pulled out of the cage by clinging to the squeeze-back apparatus and screaming. The training was discontinued.

In spite of the fact that all six juvenile candidates permitted capture in the screening test and did not attempt to bite the trainer, the majority of this set did not prove to be good candidates for the training. It was hypothesized that the juveniles, unlike adults, probably have difficulty in mastering their natural fears. Their capacity to learn is thus inhibited and they refuse to cooperate. Considering the large amount of time invested in the training of juveniles and the poor yield, the net conclusion derived from this investigation was that the training of juveniles for voluntary cooperative phlebotomies should not be recommended as a routine procedure. A secondary impression was that it may be assumed that rhesus macaques younger than 15 months have not yet developed the species-characteristic aggressive defense reaction in response to physical contact with humans.

Clarke et al. (1990) have outlined a program for the training of corral-inhabiting rhesus monkeys to be collected, caged, and voluntarily submit to a phlebotomy. The monkeys (n = 14) were daily guided into the catchpen area, placed into holding cages and following (desired) defecation were transferred to a squeeze cage. They were trained to voluntarily extend a leg through an aperture and permit a saphenous venipuncture. The animals were then returned to the corral. At the end of the 4 weeks of training, all participants were voluntarily, successfully bled at the first attempt and subsequently provided samples at 2-week intervals.

While protocols are offered for the training of monkeys to be relaxed, cooperative subjects for phlebotomy, it is also recognized that monkeys have individual personalities that aid or hinder, or at least modify their response to training. Whether or not a given monkey becomes an ideal subject seems to be dependent on its personality or perhaps its adaptability. In their study of circulating reticulocytes in rhesus monkeys, Harne et al. (1945) evaluated the individual personalities of 10, mixed-sex, adult 6 to 7-year-old *Macaca mulatta*. They were trained to be removed from their individual cages and climb upon a nearby table. Once on the table they were covered with a restraining net. The training was conducted for several months prior to experimentation. All animals were considered to be successfully trained and learned to climb on the table and wait to be covered by the net. Eventually, they would accept food treats and sit quietly while being handled. None gave indication of objection to the tests. The investigation involved taking blood daily from a cut in the auricle and on occasion 5 mL quantities of blood from superficial veins of the legs or under nembutal anesthesia 5–50 mL of blood from the femoral vein. The investigators specifically indicated while all subjects were viewed as successfully trained "each responded in a different way to the total treatment." Each monkey was assessed on its reaction to training and its individual characteristics. Examples of the descriptions of some of the monkeys are as follows: One monkey was identified by the fact that she would nervously assist in the placement of the net. Nevertheless, the same monkey never became tame. She was evaluated as an excellent subject while her reaction to training was considered poor. Another female was accorded a fair rating for her reaction to the training and further described as ugly, mean, and thievish. A third female was considered as having an excellent reaction to training and also viewed as a good subject, bossy, and amorous. One female that was cited having a fair reaction to training was further described as responsive but did things her own way. A male member of the cohort was noted as displaying an excellent response to its training and further characterized as a good subject as well as vain, thievish, and very playful. Another male who earned an excellent rating for the reaction to training was further described a good subject along with the observation that he "would be boss of colony." *Macaca mulatta* has been described as a species that demonstrates a primarily aggressive behavior toward humans.

Considering the amount of information that has been garnered regarding the scientific benefits of in-home-cage phlebotomy and the training of monkeys for voluntary, cooperative unstressed venisection it would seem logical to compare erythrograms derived from monkeys under these ideal conditions and compare them with erythrograms derived from the very same subjects following the administration of ketamine. It would help define what effects ketamine has on the erythroid profile of an unperturbed monkey. A study with this experimental design has apparently yet to be reported.

CHAPTER 5

Modern Hematology

Hematology has experienced a particularly productive period of exploration and clinical application during the past half-century. Subjects such as cellular origins, *in vitro* and *in vivo* cell culture, the identification and genetic recombinative production of hemopoietic factors, cell surface structure, and transplantation have been addressed with significant achievement. Among the most notable accomplishments has been the verification of the existence of the pluripotent hematopoietic stem cell (PHSC, HSC), its isolation, and harvest from the bone marrow and blood. Its quality of self-renewal has been experimentally established. The hematopoietic stem cell has been recognized on the basis of its unique identifying cell surface molecule, CD34+ (cluster of differentiation, CD), and its descendants have also been similarly classified on the basis of their specific individual cell surface clusters of differentiation (CD). The hematopoietic stem cell and the progenitor cells it generates are normally present in hemopoietic bone marrow but are not morphologically identifiable in standard preparations such as Wright-stained dry film smears. They are identified immunologically on the basis of their cell surface markers. The hemopoietic stem cell's cascades of progeny are not routinely morphologically recognizable until they have differentiated to the level of the pronormoblast, myeloblast, megakaryoblast, and so on. The present-day nomenclature of the hemopoietic stem cell's descendants continues to employ laboratory operational titles to signify them (such as CFU-GEMM, colony-forming unit-granulocyte erythrocyte monocyte megakaryocyte, the stem cell for the myeloid lines). In regard to the erythroid lineage specifically, its developmental ancestry is the following. The hemopoietic stem cell is envisioned as mitotically giving rise to two identical cells. One of the daughter cells persists as a hemopoietic stem cell (self-renewal) while the other enters hemopoiesis by evolving into a stem cell for the myeloid lineage, or alternatively into the stem cell for the lymphoid cells. If the cell assumes the myeloid pathway, it evolves into the CFU-GEMM. Depending on the microenvironment and ambient hemopoietic factors, this cell enters the maturational lineages of specific myeloid cells (e.g., the erythroid, neutrophilic, megakaryocytic cell lines). If the erythroid lineage is entered, the next progenitor is termed as the burst-forming unit-erythroid (BFU-E) (so-called because of its rapid and robust proliferation in cell culture systems), which in turn evolves into the further developed colony-forming unit-erythroid (CFU-E). The latter cell

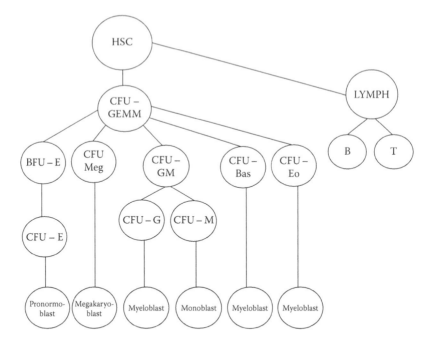

Figure 10   The hematopoietic stem cell (HSC) and its relationship with its progeny.

proliferates and differentiates into the pronormoblast, the most immature erythroid cell that is morphologically recognizable in routine preparations as a cell solely committed to the erythrocytic lineage (Figure 10).

Erythropoietin, the best known and most widely clinically employed of all hematopoietic stimulatory factors, is recognized as the single, most specific hemopoietic factor that controls the production of erythrocytes. It is produced by interstitial peritubular fibroblasts in the cortex of the kidney. This agent stimulates the proliferative capability of BFU-E, has its greatest impact on the subsequent stage CFU-E, and also targets the pronormoblasts and basophilic normoblasts.

The hemopoietic cells of bone marrow from *Macaca mulatta* (and other species) can be cultured *in vitro*. The pluripotent stem cell along with the immediate descendants can be maintained and will proliferate, differentiate, and generate progeny. BFU-E and CFU-GM assays are representative of what can be accomplished (Watanabe et al. 1990; Hillyer et al. 1993; Braun et al. 2013). The relative quantitative frequencies of CD34+ cells among the population of mononuclear cells in the blood and marrow of fetal, newborn, infant, adult, and aged rhesus monkeys have been assessed by immunostaining and flow cytometry (Chang et al. 2005). The greatest numbers of CD34+ cells present in the blood were observed in the circulation of 90 days gestational age fetuses (~5%). Older, 120 days of gestation fetuses had ~2.5%. Thereafter, the level remained at about 1% with the lowest being obtained in aged monkeys (~0.6%). In contrast, as would be hypothesized, this cell was consistently more plentiful in bone marrow (except at 90 days gestation). A steady increase in bone marrow CD34+ cells was seen afterward from 90 days gestation (~3.5%) and it peaked at ~5% in the newborn. A recession then ensued and persisted into old age (1.5%) (Figure 11). Mononuclear cells obtained from the blood and bone marrow of the same group of various aged monkeys were cultured and assayed for BFU-E cells. The quantity of BFU-E derived from blood diminished over time. The greatest yield was obtained from the blood of subjects of 90 gestational days. From then on, the frequency diminished until at 140 gestational days it was minimal and further receded to nil generation from the samples obtained throughout the rest of the life span. The marrow was consistently a richer source for cultures that generated this erythroid progenitor. At 90 days gestation, the quantitative yield was about three times that of blood. The level then dropped in the 140-gestational day fetus to slightly higher than the level obtained from blood at that time. From then on, the generation of the cell in the cultures steadily increased in parallel with age with samples from aged monkeys yielding numbers of BFU-E roughly equal to that harvested from bone marrow cultures from 90 gestational days fetuses (Figure 11).

The pluripotent hemopoietic stem cells of man and the rhesus macaque display an intimate homogeneity (Rosenzweig et al. 2001). A remarkable phenotypic and functional similarity is identifiable in rhesus and human hemopoietic progenitor cells. The frequency of CD34+ cells in macaque bone marrow, for example, correlates well with what has been established for human bone marrow. The human hematopoietic stem cell displays considerable heterogeneity. Thus, it is appropriate to state that the cells bearing the CD34+ marker are a pool of cells among which are cells possessing pluripotential hemopoietic capability as well as the property of self-renewal. However, some other cells that display this label are already further evolved. In some cases at least, they are recognizable by the other cell markers that they display in addition to CD34+ marker. This characteristic is also expressed by the rhesus monkey hemopoietic stem cell. Thus, cells with the phenotype CD34+71+ in the rhesus monkey will generate in culture cells demonstrating erythroid differentiation just as observed in man (Rosenzweig et al. 2001). In addition, cultures of CD34+61+ cells and CD34+64+ cells give rise to megakaryocytes and myeloid progeny, respectively, both in man and in *Macaca mulatta*. Rhesus marrow demonstrates CD34+38− (primitive) and CD34+38+ (lineage-committed) subsets of CD34+. Rhesus CD34+38− cells are approximately 150-fold enriched for long-term culture-initiating cells compared with unfractionated rhesus bone marrow and occur at a frequency similar to that of man (Rosenzweig et al. 2001).

The bone marrow cells of the rhesus monkey can reconstitute the hemopoietic system of lethally irradiated autologous recipients. This well-attested phenomenon documents, among other imperatives, the occurrence of the hemopoietic stem in this locus. It also implies that this system could be utilized as an assay for the presence of the hematopoietic stem cell in biological samples. The HSC (hematopoietic stem cell) and hemopoietic progenitor cells of rhesus bone marrow have physical separation characteristics similar to the cell populations found in human bone marrow. A high degree of cross-reactivity of antihuman monoclonal antibodies with cell surface antigens on rhesus hematopoietic cells allows the rhesus monkey to serve as a nonhuman model for the study of human hemopoiesis. It also serves in the

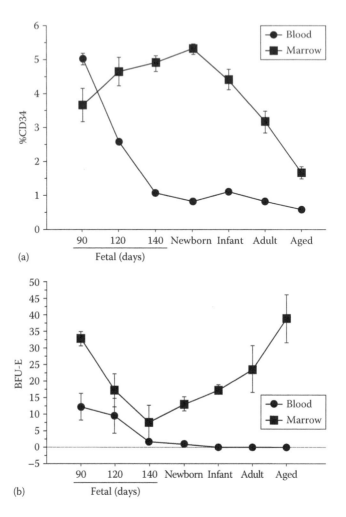

Figure 11 *Macaca mulatta.* (a) A depiction of the occurrence of CD34+ cells in the blood and bone marrow of the rhesus monkey (*Macaca mulatta*) at various ages. Mononuclear cells were obtained from the blood, immunologically stained, and quantitated by flow cytometry. It can be seen that the cells are relatively numerous in the blood during early fetal life but thereafter is present in very low levels. The bone marrow has a higher concentration of the cells than the blood. The frequency peaks in the newborn and then gradually diminishes through infancy, adult, and aged life. (b) The occurrence of BFU-E derived from the culture of mononuclear cells obtained from the blood and bone marrow of rhesus monkeys at various ages. The blood is seen to generate small numbers in the fetus and essentially none thereafter. The marrow is a far richer source for the generation of this erythroid progenitor. The curve is V-shaped with the highest yields of cultured cells derived from samples of the fetus at 90 days of gestation and from the aged monkey. (From Chang, C. et al., *Pediatr. Res.*, 58, 315, 2005.)

testing of gene therapies for diseases of the hemopoietic system (Monroy et al. 1986; Van Beusechem and Valerio 1996).

The close relationship of human and rhesus hemopoiesis is illustrated by the positive results derived from the continuous subcutaneous infusion of the recombinant human hemopoietic colony-stimulating factor (CSF) granulocyte-macrophage colony-stimulating factor (GM-CSF) into a healthy and also into a severely debilitated simian-retrovirus-infected, immunocompromised, pancytopenic *Macaca mulatta* (Donahue et al. 1986). Both subjects yielded the same response from a 7-day infusion delivered via a surgically implanted osmotic pump. The debilitated animal increased its circulating leukocyte count from 1,600 cells to 43,000 μL^{-1} at 2 days after the end of the administration of the hemopoietic factor while the healthy rhesus monkey displayed an increase of its leukocytes from 6,500 to 64,300 μL^{-1} during the same interval. Both monkeys' white cell counts returned to their original levels following cessation of the infusion. In the case of the pancytopenic monkey, a dramatic reticulocytosis and a prominent increase in the level of erythropoietin were elicited by the infusion of GM-CSF. The normal monkey had a much lower but substantial increase in reticulocytes. The continuous infusion of the latter hemopoietic factor similarly caused the circulating leukocyte counts to increase in *Macaca fascicularis*. One subject's femoral marrow aspirate was analyzed on day 7 of treatment. It revealed that the marrow had become hyperplastic during the infusion.

Consistent with expectations, human peripheral blood hematopoietic stem cells have the capability to engraft in *Macaca mulatta* fetuses *in utero* during the fetal immature immunologic period (prior to 80 days gestation) (Tarantal et al. 2000). Such chimeric infants are healthy at birth. Progenitor cell assays of blood and bone marrow of four engrafted infants were conducted up to 6 months of age and were positive for donor CFU-E and CFU-GM.

The isolation and characterization of nonhuman primate embryonic stem cells (ES) derived from monkey blastocysts have provided an experimental model with characteristics close to human biology. Stable embryonic stem cell lines have been developed for *Macaca mulatta* and *Macaca fascicularis* and are available for research (Umeda et al. 2004; Li et al. 2006). The rhesus monkey was the first, and the cynomolgus monkey was the next to be utilized in this context. The differentiation of nonhuman primate ES cells *in vitro* to form hematopoietic precursors has been intensively investigated. Such inquiries have documented the emergence of hematopoietic stem cells and their development, revealing implications closely related to human erythropoiesis. It has been observed, among other findings, that nonhuman primate ES cells possess the capability to differentiate into hemopoietic progenitor cells whose progeny belong to the primitive embryonic and definitive erythroid lineages (as well as myeloid and lymphoid lines).

CHAPTER 6

Bone Marrow
Erythropoiesis

Dry, thin-film bone marrow smears derived from aspiration biopsies of hemopoietic (red) bone marrow stained with one of the Romanowsky-type dyes (e.g., Wright's stain) have proven to be inordinately useful in the cytologic analysis of hemopoietic bone marrow. The staining of dry film smears of blood and bone marrow with a Romanowsky-type mixture, that is, eosin, methylene blue, and oxidation products of the latter dye generically termed azure stain, all dissolved in methyl alcohol has been a standard practice for more than a century and continues to remain as the single, universally accepted, most useful technique of visualizing erythrocytes and normoblasts (as well as immature and mature leukocytes) under bright field microscopy. The various Romanowsky stains are typically identified by an investigator's name who espoused certain techniques in the production of the stain that enhanced its effectiveness, predictability, and particularly the process employed in the oxidation ("polychroming") of a portion of the methylene blue. The latter step generates additional new dye products such as methylene azure that thereby add an additional color(s) to the dye mixture. Thus, Wright, Giemsa, Leishman, Jenner, May-Grunwald, and their combinations are all recognized Romanowsky stains. In all instances regardless of which Romanowsky-type stain, mixture thereof, or laboratory modification is enlisted, when an ideal result is obtained in staining a blood/marrow smear/imprint the end products of all approaches are comparable.

The morphology of the developing erythroid cells of *Macaca mulatta* is typically described as equivalent to that observed in man. In fact, the developing red cells of the monkey so closely resemble those seen in man that descriptions of the monkey's cells are said to offer little new information in regard to erythrogenesis. The comparisons of measurements of the diameters of developing normoblasts in monkey and human bone marrow have proved less convincing in establishing that this non-human primate's normoblasts are smaller than those of man than overall inspections of smeared cells in this tissue. The persistent impression, therefore, is that the normoblasts of monkeys may be slightly smaller than the normoblasts of man (Sundberg et al. 1952). The diameter of mature *Macaca mulatta* erythrocytes is decidedly less than that of the human counterpart, monkey 6.6–7.2 µm versus man 7.3–7.9 µm, per Haden-Hausser erythrocytometer (Winkle 1951; Sundberg et al. 1952).

The most immature progenitor of the erythrocyte identifiable under Romanowsky-type dye staining in dry film smears is the pronormoblast (Figure 12). It is the largest cell of the erythroid lineage. Its characteristics include a very large round nucleus that occupies the vast majority of the spherical cell. The chromatin pattern is the most delicate of all members of the cell line and consists of delicate strands of chromatin (euchromatin). Nucleoli are typically present. The cytoplasm is a narrow basophilic homogeneous band whose basophilia is attributable to ribonucleoprotein, the synthetic machinery for hemoglobin. Tiny unstained, granular, or rodlike white specs

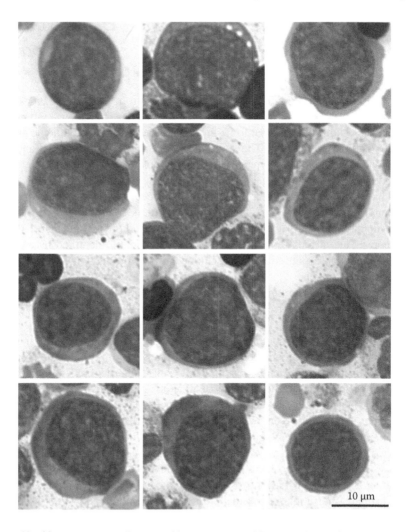

10 μm

Figure 12 *Macaca mulatta*. Pronormoblasts and basophilic normoblasts from bone marrow smear of *Macaca mulatta*. The top two horizontal rows and the first two cells of the third row are considered pronormoblasts. The remaining cells are basophilic normoblasts. Figure 13 on the adjacent page presents the continued development of normoblasts. (Wright stain.)

are often seen in the cytoplasm and are for the most part mitochondria. As in man's pronormoblasts, an unstained relatively large, prominent white area in the cytoplasm may be identified in some cells. This negative-staining region is due to the localization of the Golgi apparatus at this site.

The basophilic normoblast is the next recognized stage in the differentiation of the red cell's antecedent (Figure 12). It is similar to the pronormoblast but somewhat smaller and the chromatin pattern is coarser. Nucleoli may or may not be identifiable

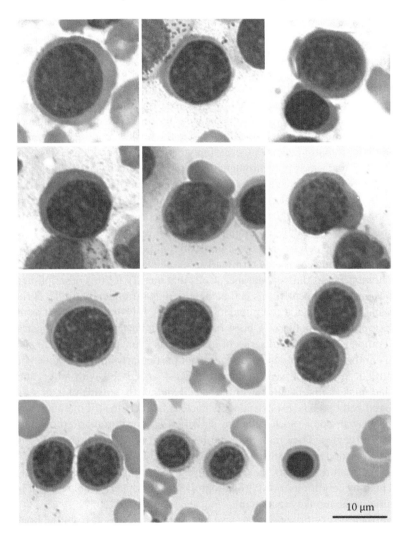

Figure 13 *Macaca mulatta*. Polychromatophilic and orthochromic normoblasts from bone marrow smear of *Macaca mulatta*. The most immature (earliest) polychromatophilic normoblasts are present in the top two horizontal rows of cells. The further developed cells are in the lower two rows. The last cell in the fourth row is an orthochromic normoblast. (Wright stain.)

and if present are smaller. The intensity of the cytoplasmic basophilia is increased and is probably this cell's most identifiable feature. The latter is considered to reflect an increased quantity of ribosomes for the synthesis of hemoglobin.

The following step in this maturational process is the polychromatophilic (poly-chromatic) normoblast (Figure 13). This cell is by far the most numerous of all normoblasts (Table 8). It is primarily identifiable by its cytoplasm that exhibits dif-fuse composite staining by both the (basic) methylene blue and (acidic) eosin of the Romanowsky-type dye mixtures. Thus, any normoblast that shows a diminution of its basophilia (blueness) because of its inclusion of eosin (thus modifying the color of the cytoplasm) is recognized as a polychromatophilic normoblast. Further, as the amount of ribosomes diminish in the course of cellular differentiation (thereby less-ening the basophilia) coupled with the accumulation of ever-increasing amounts of hemoglobin, the color of the cytoplasm becomes redder and redder while giving evi-dence of still staining with some methylene blue. The simultaneous affinity for the variable amounts of these two stains is the basis for the cell's identification as a poly-chromatophilic normoblast. On the other hand, the nucleus continues to decrease in size and the chromatin becomes increasingly clumped and condensed. Any normo-blast whose cytoplasm is polychromatic but whose nucleus has not condensed to the ultimate pre-extrusion stage is termed a polychromatophilic normoblast.

The final stage of development of the maturing, still-nucleated erythrocyte pre-cursor is the orthochromic (orthochromatic, normochromic) normoblast (Figure 13). The cytoplasm is distinctly eosinophilic but may yet display a subtle component blue that causes the cell to display a cytoplasm that is not exactly as red as a mature erythrocyte. The nucleus is now terminally clumped and reduced in size or may have a solid, homogeneous dark appearance. Cells that qualify for this description are not numerous and consequently, this cell is identified as infrequently as 2% of the mar-row nucleated hemopoietic cell population (Table 8). It is thus seen about as often as is the basophilic normoblast. A bone marrow smear that fortuitously illustrates an almost complete maturational series of normoblasts is presented in Figure 14.

Cells in the process of denucleation are not particularly commonly seen but may be identified. When observed, the usual image is a red cell with an eccentrically positioned nucleus that is partly within the cell and partly outside of it.

The red cell that has just extruded its nucleus is likely to still maintain a slightly less red, grayish color. Such cells are classified as polychromatic (polychromato-philic) erythrocytes. They remain in the bone marrow for a short while (2 days in man) and lose enough of the residual ribosomes to acquire the normal so-called brick red color of mature erythrocytes. Most investigators are in agreement that under normal healthy conditions, polychromatophilic erythrocytes are not released into the circulation in the monkey (and man).

Nevertheless, erythrocytes that have just been just released from the bone mar-row still retain a minimal amount of ribonucleoprotein but have the appearance of older cells in circulation that no longer have this cytoplasmic component. The resid-ual ribonucleoprotein is lost from the cell during the first day it is in the circulating blood. The identification of the cells that manifest the minimal ribonucleoprotein is based on the reticulocyte staining technique. That is, they are stained supravitally

Table 8 Bone Marrow Erythroid Differential Counts of *Macaca mulatta*

	Site	Age	Sex, n =	ProNbl	BasNbl	Polych	Orth	Mean Nbl	Range
Suarez et al. (1942, 1943)[a]	Stern	Infant	?, 4	0.4	1.3	5.8	12.3	19.8	10.5–31.0
		Young	♂ and ♀, 8	0.2	0.9	4.8	10.8	16.7	10.6–21.0
		Adult	♂ and ♀, 28	0.2	1.0	4.4	10.3	15.8	3.6–27.4
Winkle (1951)[b]	Ilium	10–15 mo	♂, 30	0.4	1.4	22.2	0.8	24.8	10.4–36.2
		10–15 mo	♀, 16	0.5	1.5	23.9	1.0	26.9	13.0–35.2
		10–15 mo	♂ and ♀, 8	0.4	1.3	24.8	0.4	26.9	22.4–36.8
Porter et al. (1962)[c]	Tibia	10–15 mo	?, 10	—	—	—	—	24.8	8.4–38.8
Vitamin E deficient, 17–28 mo later	Tibia	27–43 mo	?, 5	—	—	—	—	38.8	26.8–55.0
Usacheva et al. (1963)[d]	—	1–2 years	♂ and ♀, 14	0.9	1.6	26.7	1.7	30.9	10.9–38.2
Cohen (1953)[e]	Hum	Young	♂ and ♀, 25	0.02	0.6	33.3	—	33.9	14.4–51.8
Glomski et al. (1982)[f]	Tibia	Young	♂, 5	—	—	—	—	27.2*	19.4–35.5
Methylcellulose IP	Tibia	Young	♂, 5	—	—	—	—	38.7*	27.4–45.8
Switzer (1967a)[g]	Ischium	Adult	♂, 5	0.4	4.6	35.8	—	40.7	30.9–48.0 ♂ and ♀
		Adult	♀, 20	0.5	5.7	32.6	—	38.7	30.9–48.0 ♂ and ♀
Stasney et al. (1936)[h]	Stern	Adult	?, 6	—	—	—	—	40.8	—
	Rib	Adult	?, 6	—	—	—	—	42.1	—
	Vert	Adult	?, 6	—	—	—	—	40.9	—
	Femur	Adult	?, 6	—	—	—	—	45.2	—
	Tibia	Adult	?, 6	—	—	—	—	49.7	—

(Continued)

Table 8 (Continued)   Bone Marrow Erythroid Differential Counts of *Macaca mulatta*

	Site	Age	Sex, n =	ProNbl	BasNbl	Polych	Orth	Mean Nbl	Range
Huser (1970)	Ilium	Adult	♂ and ♀;?	3.0	—	39.9[x]	—	42.9	33.8–51.7
Kusova and Dikovenko (1961)[i]	Tibia	?	♂,?	1.1	4.7	19.9	1.6	27.3	9.0–52.8

Quantities of normoblasts are expressed as percent of total nucleated mature and immature hemopoietic cells observed in Romanowsky-type dye-stained dry film smears. Asterisks indicate a significant difference between compared paired values. Exponent[x] indicates the aggregate of basophilic, polychromatophilic, and orthochromic normoblasts. Stern = sternum; Hum = humerus; Vert = vertebra.

Values are mean values. ProNbl, percentage of pronormoblasts; BasNbl, percentage of basophilic normoblasts; Polych, percentage of polychromatophilic normoblasts; Orth, percentage of orthochromatophilic/orthochromic normoblasts; Mean Nbl, mean percentage of total normoblasts of differential counts; Range, range of total normoblasts identified in differential counts. Stern, sternum; hum, humerus; vert, vertebra.

a Terminology was modified from reported megaloblasts, early erythroblasts, late erythroblasts, and normoblasts.

b The sex was not recorded for the group of eight subjects. The biopsy site was just inferior or actually on the iliac crest.

c At the start of the investigation, the monkeys weighed 1.5–2.5 kg. This is the same weight range of the monkeys studied by Winkle (1951) that were considered on the basis of dental evaluation to be 10–15 months of age. The subjects of the present investigation were consequently considered to also be 10–15 months old. The sex of the subjects was not reported. Tibial aspiration biopsies were performed on 10 subjects (a total of 14 determinations). After 17–28 months on a vitamin E–deficient diet, the monkeys developed an anemia (as well as other signs of vitamin E deficiency; e.g., muscular dystrophy and creatinuria). Tibial aspiration biopsies (n = 5) of the monkeys at this time revealed a moderate erythroid hyperplasia. Many of the erythroid precursors were multinucleated (range 12%–34%). All stages of erythroid maturation were involved having up to and including 4 nuclei. Their chromatin did not have the normal appearance but tended to be homogeneous. The morphology was not megaloblastic.

d The terminology used for the erythroid precursors was proerythroblasts, basophilic macroblasts, basophilic normoblasts, polychromatophilic normoblasts, and oxyphilic normoblasts. The basophilic and polychromatophilic normoblasts were combined in the present table into polychromatophilic normoblasts.

e The nomenclature for the erythroid precursors was megaloblast, early erythroblast, late erythroblast, and normoblast. These titles were converted into the classification used in the present table except the late erythroblast and normoblast were combined and listed as polychromatophilic normoblasts to avoid a disproportionately large value for the orthochromic normoblasts. Bone marrow aspirations were obtained from the greater tuberosity of the humerus. The weights of the monkeys were 2.7–5.5 kg. Judging from the weights of the subjects they were assumed to be relatively young monkeys.

f The weights of the monkeys ranged from 2.5 to 5 kg. The monkeys were placed on a regimen of biweekly intraperitoneal injections of 2.5% aqueous solution of methylcellulose for 14 weeks for a total dosage of 5.6 g/kg body weight. Tibial aspiration bone marrow biopsies were performed prior to the initiation of administration of methylcellulose and at autopsy 3 months after completion of the series of injections. The asterisks indicate a significant difference in the pre- and posttreatment levels of normoblasts in the bone marrow.

g The marrow was aspirated from the ischial tuberosity. Weight of males: 5.4–7.6 kg, mean age 5.8 years old; weight of females: 3.7–8.2 kg, mean age 8.2 years old and at least 60 days postpartum. In the original report, the members of the erythroid lineage were classified as rubriblasts, prorubricytes, basophil rubricytes, polychromatophil rubricytes, and metarubricytes. In order to conveniently accommodate the cells to the nomenclature utilized in the present table, the rubriblasts were classified as pronormoblasts, the prorubricytes and basophil rubricytes were listed as basophilic normoblasts, and the polychromatophil rubricytes along with the metarubricytes were grouped as polychromatophilic normoblasts.

h Imprint preparations ("smears") were prepared from the bone marrow of five different locations of the body of six adult monkeys that had been residents of the laboratory for 3 years. Differential counts were determined on the imprints.

i Kusova and Dikovenko state "Since the haemopoietic values in males are more stable than in females, data for males are presented."

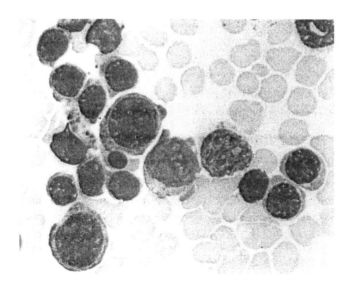

Figure 14 *Macaca mulatta.* A cluster of normoblasts in a bone marrow smear of a *Macaca mulatta* monkey. This smear fortuitously presents an almost complete maturational series of normoblasts. The central cell with relatively prominent nucleoli, a delicate chromatin pattern, and lightly stained cytoplasm is a probable pronormoblast. The slightly smaller cell on the right is a damaged, unrecognized cell. On the adjacent left and above the pronormoblast is a basophilic normoblast. A similarly large spherical cell positioned in the lowest area of the field presenting a well-defined narrow rim of less basophilic cytoplasm and a nucleus with a somewhat clumped pattern of its chromatin is identified as a late basophilic or an early polychromatophilic normoblast. Two considerably smaller, medium-sized cells located along the right border of the photomicrograph are older, more mature, easily recognizable polychromatic forms. Four further developed cells that are additionally reduced in dimension are located immediately superior and slightly left of the basophilic normoblast. These are late polychromatophilic normoblasts. A morphologically ideal normochromic/orthochromic normoblast is not observed in the field, although an imperfectly and incompletely displayed normochromic normoblast, with an appropriate, dense-staining small pyknotic nucleus, is seen immediately below the basophilic normoblast. (Wright stain; oil immersion magnification.) (From Winkle, V.A., The study of bone marrow and blood of normal laboratory monkeys, Thesis, University of Minnesota, Minneapolis, MN, 1951.)

with a basic stain such as new methylene blue or brilliant cresyl blue. The cells that have residual ribosomes demonstrate a blue precipitate that is seen under routine light microscopy. These cells are termed reticulocytes.

The progressive decrease in the size of erythroid cells as they undergo maturation is readily identifiable on examination of a bone marrow smear. Figures 12 and 13 demonstrate this somewhat stepwise diminution. All cells in these two figures were photomicrographed under the same magnification. The youngest cells (pro- and basophilic normoblasts) by and large occupy most of the equally allotted spaces while the subsequent polychromatophilic normoblasts are smaller and have considerably more excess space surrounding them. In man, the pronormoblast has a cellular volume of 900 fL while the end cell, the mature erythrocyte, is reduced to 95 fL.

A roughly comparable equivalent reduction is seen in the rhesus monkey. While its pronormoblast is approximately 10%–15% smaller than that of man, its red cell is also smaller in the range of 75 fL.

The consideration of this marked change is that cell size brings to mind the huge amount of the protein hemoglobin that is synthesized by the erythrocytic progenitors and the increasing concentration of this product in successively smaller cells. While hemoglobin is not tinctorially identifiable in the pronormoblast under Romanowsky staining, it is nevertheless synthesized at this early stage. According to Lajtha's (1965) model of the "erythron" (the aggregate population of red cell precursors in an individual), man's basophilic normoblast, for example, has a mean content of 15 pg of Hb. The presence of hemoglobin in this progenitor is cytochemically documented by the benzidine peroxidase reaction. Hemoglobin is continually produced and accumulates in the cell as recognized by polychromatic and orthochromic staining of the cytoplasm. In addition, hemoglobin is considered to be synthesized for a short period following enucleation by the remaining ribosomes of the reticulocyte. The accumulation and concentration of hemoglobin increase to the point, as indicated by the MCHC, that it represents ~32% of the cell.

The pronormoblast is the least numerous cell of the erythrocytic lineage (Table 8). The basophilic normoblast is slightly more numerous while the most plentiful member of the line is the polychromatophilic normoblast. This progressive amplification in numbers is due to the fact that the erythrocytic precursors proliferate while undergoing maturation. In man, one pronormoblast generates on average 16 erythrocytes (i.e., 3–5 sets of mitotic events). Failure to attain this yield is considered ineffective erythropoiesis. The number of descendants for the rhesus pronormoblast is not readily identified but is unlikely to be less since the erythrocyte count of the monkey is higher than that of man and also because the macaque's red cell has a greater rate of turnover (a shorter life span, ~100 days for the monkey versus 120 days for man).

Mitotic figures of normoblasts are relatively numerous in bone marrow smears and are readily identifiable (Figure 15). Another cytologic feature of erythroid progenitor development is that normoblasts in the polychromatophilic and orthochromic stages of maturation frequently display a long thin strand of cytoplasm between two daughter cells (Figure 15). It is identifiable in the monkey. This cytoplasmic bridge is well recognized while the function of this extended connecting strand is not fully understood. It is not observed between recently mitosed leukocytes. This filamentous structure has been photomicrographically documented in human normal and pathological bone marrow (Kobayashi et al. 1990). In the latter, workers' experience electron microscopy revealed that small amounts of microtubules were contained therein. Approximately, in the middle of its length the plasma membrane bulged but did not contain an electron-dense organellar structure (the midbody) observed at the telophase of animal cells at cytokinesis (Jones 1969; Alberts et al. 1983). The partners of individual pairs of linked cells were typically at the same level of differentiation. Kobayashi et al. reported that a mean 8% of the normoblasts in 33 normal marrows displayed this structure. Conversely, patients with autoimmune hemolytic anemia, aplastic anemia, paroxysmal nocturnal hemolytic anemia, eythroleukemia, myelodysplastic syndrome, and refractory anemia with excess blasts usually

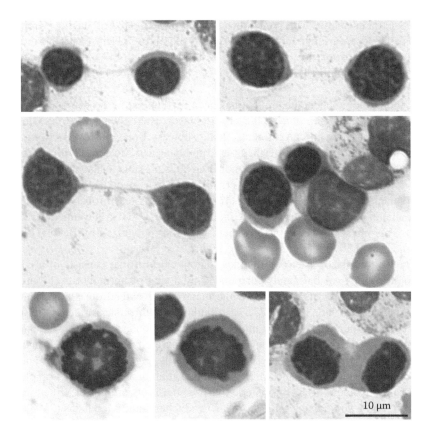

Figure 15 *Macaca mulatta*. Cells obtained from bone marrow smear of *Macaca mulatta*. Top row and first frame of middle row: Three pairs of normoblasts that have recently undergone mitosis displaying a long, thin extension of cytoplasm that continues to link the two daughter cells. Middle row, second frame: A cluster of three cells composed of two late polychromatophilic normoblasts and an adjacent lymphocyte. The lymphocyte and the polychromatophilic normoblast may at times have a "lookalike appearance" requiring morphologic consideration to identify them correctly. Bottom row: Mitotic figures of normoblasts. (Wright stain.)

displayed about one half or less of that frequency. It was proposed that this cytoplasmic bridge was a normal consequence of, and possibly essential for, the normal proliferation of normoblasts.

Lymphocytes are a standard component of hemopoietic bone marrow. In the rhesus monkey, they comprise a population that is numerically about 75% as large as that of the normoblasts (Winkle 1951, Figure 15). The small lymphocyte (by far the most numerous form of this cell) and the polychromatophilic normoblast have at times a similar appearance requiring some morphologic analysis to insure their correct identification. The lymphocyte presents a dense smudged nucleus with little morphologic detail, while the normoblast nucleus has a "checkerboard" pattern that is described as having relatively demarcated clumps of chromatin separated by areas termed parachromatin. The lymphocytes are typically cited as having cytoplasm that is stained a pure blue color while that of the early polychromatic normoblast is a homogeneous mixture of combined eosinophilic and blue staining (due to the presence of acidophilic hemoglobin and basophilic ribosomes) yielding a color that is not pure blue.

A morphologic phenomenon that has been recognized in the hemopoietic bone marrow of *Macaca mulatta* is the erythroblastic islet (Switzer 1967a; Glomski personal observation in one immature male subject). Bone marrow sections from rhesus monkeys have been described by Krumbhaar and Musser (1923) as presenting "at times distinct clumping of normoblasts." It is possible that some of these may have been erythroblastic islets. This organization is a specialized cellular micro-environmental niche in which mammalian normoblasts (erythroblasts) proliferate, differentiate, and enucleate (Chasis and Mohandas 2008; Manwani and Bieker 2008). This physical cellular arrangement was first recognized by Bessis (1958) a little more than 50 years ago.

The erythroblastic islet is a distinct anatomical unit that consists of a central macrophage (or two) intimately surrounded by a ring of normoblasts (Figures 16 and 17). Cytoplasmic pseudopods of the macrophage extend out among the normoblasts (Bessis 1966, 1973, 1977). In stimulated erythropoiesis, the central macrophage is surrounded by one or more layers of normoblasts in different stages of maturation. The central macrophage has been described as a nurse cell. Progenitor niches such as this have also described other cells (e.g., the Sertoli cell and the developing spermatocytes and spermatozoa that it embraces). Though accepted as a normal constituent of erythropoietic marrow, the islets are not identified as frequently as one would expect. Bessis (1977) and Chasis and Mohandas (2008) have indicated that this is due to the fact that the central macrophage and its long projections are very fragile and are disrupted upon smear preparation. The cytoplasmic processes break up into several fragments that round up. The latter bits of cytoplasm are usually overlooked in bone marrow smears. Single normoblasts are often identified in close association with a macrophage and may be assumed to perhaps represent the remains of a disorganized islet. Bessis (1973, 1977) has also described and illustrated fragments of cytoplasm positioned adjacent to normoblasts and has cited these adherent cytoplasmic masses as remnants of the central macrophage. The occurrence of the erythroblastic islets was confirmed in 3D scale models of microscopically sectioned rat bone marrow constructed by Mohandas and Prenant (1978). They were distributed throughout the marrow space. From a comparative hematologic aspect, it is noteworthy that the islets are apparently not observed in avian or poikilothermic (cold blooded) vertebrates' bone marrow (Bessis 1977; Sorrell and Weiss 1982; Socolovsky 2013). It is assumed that this special anatomical relationship between the macrophage and the erythroblasts did not develop since erythrocytes in these species do not extrude their nuclei and the entire intact red cells circulate in their blood.

The relationships that the central macrophage of the islet maintains with attendant normoblasts have been of recent interest. It has been documented that erythrocyte progenitors display autonomous differentiation and have the capacity to complete

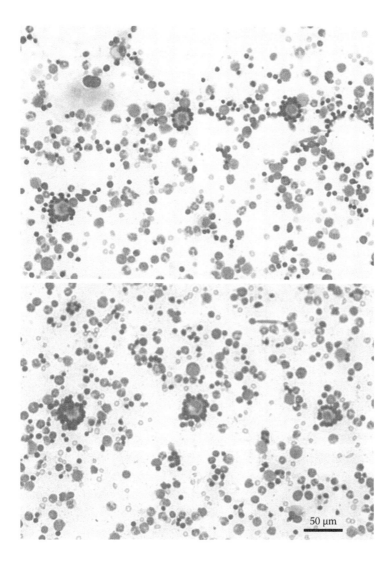

Figure 16 *Macaca mulatta*. Erythroblastic islets observed at scanning microscopic power. Wright-stained dry film smears of bone marrow from a *Macaca mulatta* monkey.

terminal differentiation *in vitro* in the presence of erythropoietin but in the absence of macrophages. However, Chasis and Mohandas (2008) have pointed out that the extent of proliferation, differentiation, and efficiency of enucleation observed *in vivo* could not be duplicated *in vitro*. This has led to further analyses of the processes conducted within the islets. The contributions of the erythroblastic islet to erythropoiesis include the following: (1) erythroblasts (normoblasts) differentiate into red cells within the erythroblastic islet niche where they are attached, (2) the macrophages amplify the normoblasts' response to erythropoietin the major regulator of

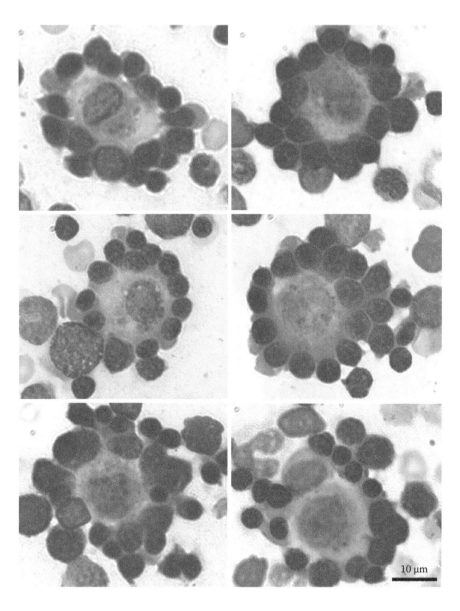

Figure 17 *Macaca mulatta*. Erythroblastic islets in dry film smear of bone marrow from *Macaca mulatta*. Phagocytosed material is identifiable in some of the macrophages. (Oil immersion magnification; Wright stain.)

red cell production by promoting normoblastic cell division, (3) the islet assists in enucleation and iron uptake as well as holds the normoblasts in close proximity to allow reciprocal autoregulatory interactions (Socolovsky 2013).

Further advances have been made in the analysis of the functions of the erythroblastic islet. In the case of Toda et al. (2014), phenylhydrazine was injected into mice and induced a hemolytic anemia along with the generation of erythroblastic islets in the spleen of the recipient. The islets were isolated and an *in vitro* system was established to study the functions of the erythroblastic islets. The production of erythrocytes in an erythroblastic islet was described as occurring in four steps. (1) The normoblast binds to the macrophage in the center of the islet through the interaction of integrin $\alpha_4\beta_1$ on the normoblast and Vcam1 on the macrophage. (2) The normoblast matures and is accompanied by enucleation. (3) The latter separation yields a pyrenocyte and a reticulocyte. A pyrenocyte is the extruded nucleus surrounded by a plasma membrane. When the pyrenocyte separates from the reticulocyte, this action exposes phosphatidylserine which is distributed on the surface of the pyrenocyte. Protein S, which is in the serum, functions as a bridge between the pyrenocyte and macrophage by binding to the phosphatidylserine on the pyrenocytes and to MerTK (a receptor kinase) on the macrophage. (4) Engulfment of the pyrenocyte by the macrophage now occurs. The phagocytosis of the intact pyrenocyte enwrapped by plasma membrane is hypothesized to prevent the release of nuclear contents that could activate the immune system.

Another concept regarding the monkey's normoblasts merits consideration. It is unrelated to and independent of the erythroblastic islet. A potential cytoarchitectural feature of macaque red cell precursors whose existence remains to be recognized or definitively denied is a system of channels in the normoblastic nucleus. The channels were discovered in fishes' (permanently nucleated) erythrocytes by Jagoe and Welter (1995). They are identifiable in experimentally denuded ichthyan erythrocytic nuclei (devoid of plasmalemma, cytoplasm, and nuclear membrane) and observed on these nuclear masses as conspicuous apertures or lacunae on average ~100–200 nμ in diameter (as opposed to the ~80 nμ diameter of nuclear membrane pores). They are covered by the nuclear membrane and may comprise ~10% of the nuclear surface. Further, in whole cell transmission electron microscopic (TEM) sections, the "holes" are identifiable as the ends of channels within the nucleus. This organization was identified thus far in the erythrocytes of seven species of teleosts (bony fishes), for example, the largemouth bass *Micropterus salmoides* and the yellow perch *Perca flavescens*. Interestingly, their presence is also consistent with the TEM and light microscopy of red cells in an amphibian, the leopard frog *Rana pipiens*, and an avian, the chicken *Gallus domesticus* (Davies 1961). A relevant observation is that these nuclear channels are not present in ichthyan leukocytes as well as some studied non-erythroid mammalian cells. Since the structure of submammalian erythrocytes is remarkably uniform, this system of apertures is likely to be demonstrable in red cells from the representatives of other taxa (perhaps including mammalian normoblasts). The existence of this organization gives rise to the question of whether it has a relationship with or is responsible for the checkerboard pattern that serves as the signature of normoblastic nuclear morphology under bright field microscopy.

Moreover, this nuclear architectural characteristic invokes the suggestion that the macro-organization of the DNA and associated proteins in nucleated erythroid cells may be different from other vertebrate cells that are not of erythrocytic lineage. It remains to be seen whether the nucleus of the erythroid precursors of the *Macaca mulatta* presents this structural organization.

CHAPTER 7

Bone Marrow
Erythrogram

Sternal bone marrow aspirations were performed on 4 infant, 8 young, and 28 adult *Macaca mulatta* that were residents of the Santiago Island Primate Colony, Puerto Rico where environmental conditions approach the most natural habitats for these animals (Suarez et al. 1942, 1943). The monkeys were caged only a day or two prior to the experiment. Twenty-one monkeys were males and 19 were females, but the sexes of the animals in the age groups were not otherwise indicated. Dry film smears were prepared and stained with Jenner–Giemsa stain. Differential counts were obtained on all subjects. The nomenclature for the erythroid series used for this study was megaloblast (for the youngest progenitor), followed by the early erythroblast, and then late erythroblast, and finally the normoblast. It can be assumed that these four stages are equivalent to the pronormoblast, basophilic normoblast, the early polychromatophilic (polychromatic) normoblast, and the pool of cells comprised of late polychromatic and orthochromic normoblasts.

The youngest monkey of the group of four infants in Suarez et al.'s investigation was 6 months old and weighed 1.5 kg. As one would anticipate, the differential count revealed that the pronormoblast was the rarest erythroid precursor (mean 0.4% of the population of nucleated mature and immature hematogenous cells in the marrow). The basophilic normoblasts comprised a mean 1.3%, while the younger polychromatophilic forms totaled 5.8%. The group of late polychromatic and normochromic normoblasts (listed as orthochromic normoblasts in Table 8) represented a mean 12.3% of the marrow hemic cells. The mean for the total erythroid precursors was 19.8% (range 10.5%–31.0%, Table 8).

The cohort of young *Macaca mulatta* of Suarez et al. (1942, 1943) totaled eight subjects. Representatives of each level of normoblastic maturation were less frequently identified than they were in the infants' marrows. The mean total of nucleated erythroid precursors was 16.7% with a range of 10.6%–21.0% (Table 8).

Twenty-eight mature rhesus monkeys comprised the adult group (Suarez et al. 1942, 1943). The oldest age was 12 years, and the maximum weight of an individual in the group was 11.5 kg. In general, there was a further diminished frequency of the various stages of normoblasts (Table 8). The mean total occurrence of immature nucleated erythrocytic progenitors was 15.8%. The range extended from 3.6% to 27.4%.

A cytologic analysis of the hemopoietic bone marrow of normal, laboratory-housed immature *Macaca mulatta* (n = 54, estimated age per dental assessment 10–15 months, mean weight 1.5–2.5 kg, male to female ratio ~2–1) was conducted by Winkle (1951). A minimal amount of marrow was aspirated from the ilium typically just inferior to or on the iliac crest (generally 2–5 drops), a quantity just sufficient just to make direct smears. The direct dry film smears were made without the use of an anticoagulant and stained with Wright's stain. The study revealed that the sum total of normoblasts represented a mean 25.7% of the marrow hemopoietic cell population. The polychromatophilic (or polylchromatic) normoblasts were the predominantly identified erythroid precursors (mean 23.1% of the marrow nucleated hemopoietic cells). The pronormoblasts totaled a mean 0.46%; basophilic normoblasts were slightly more common (mean 1.4%), while the most mature nucleated erythroid precursors, the orthochromic normoblasts, were also sparse (mean 0.80%) (percentages of the hemopoietic cell population). This relatively large cohort of healthy immature rhesus monkeys revealed, perhaps surprisingly, a wide range in the occurrence of normoblasts in hemopoietic bone marrow. The range extended from 10% to 37% of the total hemopoietic population. The 30 males of this cohort had a mean 24.8% normoblasts, while the 16 females had 26.9% erythroid precursors (Table 8). (The sex of eight subjects was not recorded, and consequently, these monkeys were not included in this calculation.) The data were subjected to statistical analysis to establish whether there was a significant difference in the number of normoblasts in the two sexes. The subpopulation for females was as follows: $\overline{Y}_1 = 26.9\%$, s.d. = 6.7; for the male subpopulation: $\overline{Y}_2 = 24.8\%$, s.d. = 7.3. The difference in means between the two sexes per analysis was not statistically significant (P = 0.30). The level of normoblasts of the eight other members of this colony of the same age and other pertinent characteristics but of unidentified sex had the same normoblast representation of 26.9% (Table 8, Winkle 1951). Young laboratory-housed rhesus monkeys of the same weight range as the monkeys investigated by Winkle (1951) presented the same mean level of normoblasts in tibial aspirations as the males of Winkle's ileal studies (n = 10, sex not reported, Table 8, Porter et al. 1962).

Bone marrow differential counts performed on 14 healthy, single-caged mixed-sex rhesus monkeys 1–2 years old yielded a mean 30.9% erythroid precursors, ranging 10.9%–38.2% (Table 8, Usacheva and Raeva 1963). These subjects were thus about 1 year older than the monkeys evaluated by Winkle (1951).

An interesting comment made by Usacheva and Raeva (1963) is that the mean concentration of nucleated hemopoietic cells in the bone marrow (specific bone not identified) as derived from an aspiration biopsy was 101,000/mm^3, range 56,000–201,000 cells/mm^3 (determined with Goryaev's counting chamber).

Bone marrow aspiration biopsies from the greater tuberosity of the humerus from 25 healthy rhesus monkeys were assessed at the Chemical Corps Medical Laboratories, U.S. Army Chemical Center, MD (both sexes were included, weight range 2.7–5.5 kg) (Cohen 1953). This site was observed to be adaptable for repeated aspirations (e.g., as often as five times within a 2-month period). Dry film smears were prepared and stained with Wright's stain. The mean combined erythroblast (normoblast) value was

33.9%, range 14.4%–51.8%. Tibial bone marrow aspirations from five young healthy, laboratory-housed male *Macaca mulatta* of the same weight range demonstrated a mean 27.2% normoblasts (range 19.4%–35.5%) (Table 8, Glomski et al. 1982).

An investigation of the hemopoietic marrow (and peripheral blood) of healthy adult male and female *Macaca mulatta* has been conducted by Switzer (1967a, Tables 1 and 8). Five males and 20 females were assessed (laboratory-maintained, singly caged, ♂5.4–7.6 kg, mean age 5.8 years old; ♀3.7–8.2 kg, mean age 8.2 years old, and at least 60 days postpartum). The marrow was aspirated from the ischial tuberosity; direct dry film smears were made and stained with Wright's (one film) and Wright–Giemsa (one film) stains. The erythroid precursors were classified according to the alternative nomenclature based on the stem word *ruber* or *rubro*, the Latin word for red. Thus, the cells were identified as rubriblasts, the equivalent for pronormoblasts, prorubricytes and basophil rubricytes (which can be categorized as basophilic normoblasts), and polychromatophil rubricytes plus metarubricytes, which together can be considered to include the polychromatophilic normoblasts and the further developed orthochromic normoblasts. As similarly observed in the study of immature, ~1-year-old rhesus monkeys of Winkle (1951) the rubriblasts/pronormoblasts, the earliest cells committed to the erythroid series that are identifiable by staining with a Romanowsky-type dye (e.g., Wright, Giemsa stains) made up ≤1% of the marrow hemopoietic cell population in the combined male and female differential count. The next more common precursors, the prorubricytes/basophilic normoblasts, totaled 0.9%–2.6% of the marrow cells. The cells identified by Switzer as basophil rubricytes embraced 2%–6.4% of the hemic cell population. Polychromatophil rubricytes and metarubricytes comprised in aggregate a range of 21.6%–50.8% of the hemic nucleated marrow erythropoietic cells. The total mean representation of erythroid precursors in these monkeys was 40.7% and 38.7% for the males and females, respectively (Table 8). These values are considerably higher than the 25.7% obtained in the assessment of 54 immature ~1-year-old macaques described by Winkle (1951). As in the case of Winkle's immature monkeys, Switzer's (1967a) adult subjects also did not show any significant difference in the numbers of erythroid precursors in the males' and females' bone marrows.

The early study of the bone marrow of the rhesus monkey by Stasney and Higgins (1936, Table 8) (credited by Huser [1970] as the first myelograms to be performed on this species) also compared the levels of hemopoietic activity observed in various bones of this nonhuman primate. The subjects were six sex-unidentified adults that had been laboratory housed for 3 years. Immediately after presumed euthanasia, bone marrow was taken from five regions, the sternum, rib, vertebra, right femur, and right tibia. Specimens were fixed; paraffin sections were prepared and stained with Dominici stain (Figures 18 and 19). In addition, imprint ("touch") preparations were made from each region and stained with May-Grunwald-Giemsa stain. The microscopic and cytologic end results of imprints are essentially that of dry film smears, although the overall enlargement is said to be slightly reduced (Figure 20). This combination of techniques permitted the concurrent analysis of the identical tissues under two formats, that is, in sectioned material that permitted the study of

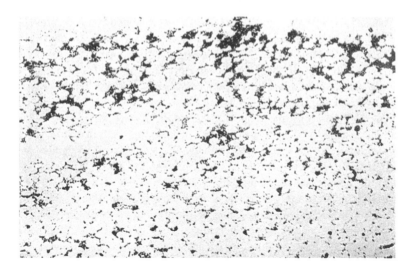

Figure 18 *Macaca mulatta*. A section of bone marrow from a tibia of an adult *Macaca mulatta*. This is an example of bone marrow that is predominantly made up of adipose cells (yellow marrow) with some scattered clusters of hemopoietic cells. The hemopoietic cells are more numerous in the upper area of the illustration and were located in the region immediately adjacent to the bone (not included in the photomicrograph). ×60. (From Stasney, J. and Higgins, G.M., *Anat. Rec.*, 67, 219, 1936.)

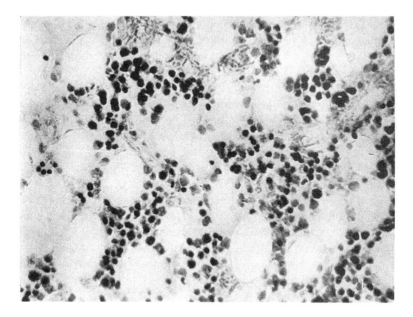

Figure 19 *Macaca mulatta*. A section of bone marrow from a tibia of an adult *Macaca mulatta* from an area that showed active hemopoiesis. (See Figure 18.) Even in this region, fat cells, seen as large spherical unstained areas, make up approximately 50% of the marrow volume. ×435. (From Stasney, J. and Higgins, G.M., *Anat. Rec.*, 67, 219, 1936.)

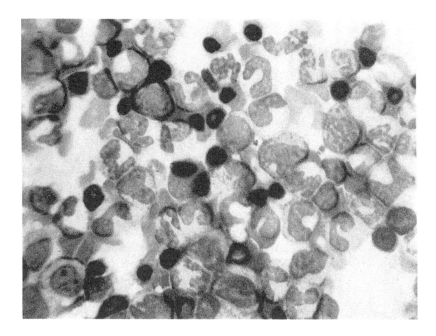

Figure 20 *Macaca mulatta*. An imprint ("touch") preparation of bone marrow from the sternum of an adult *Macaca mulatta*. Representatives of myeloid and erythroid lineages are seen. The normoblasts are the smaller, round cells with solid dark-staining nuclei. (May-Grunwald-Giemsa stain.) ×1000. (From Stasney, J. and Higgins, G.M., *Anat. Rec.*, 67, 219, 1936.)

the architecture and cellular distribution of the bone marrow and in ideal cytologic preparations that are equivalent to dry film smears (Figure 20). The imprints offered the advantage of avoiding dilution of the marrow tissue with fluid and cells from the peripheral blood as is routinely encountered with aspiration biopsies. The differential counts thereby included only cells that were present in the marrow.

The mean occurrences of normoblasts in the sternum, rib, and vertebra in the monkeys were essentially identical (40.8%, 42.1%, and 40.9%, respectively; mean 41.3%) (Stasney and Higgins 1936). The incidence of normoblasts among the nucleated hemic cells of the marrow in the long bones in these subjects was significantly higher (femur mean 45.2%, tibia mean 49.7%). The tibia was particularly fatty with hemopoietic foci scattered throughout the fatty areas. The hemopoietic cells in this bone were particularly localized in the subosseous areas of the marrow (Figure 18). It is also seen that the representation of normoblasts in the marrow of the sternum, rib, and vertebra in these mature monkeys is in the same range as that reported for the ischial aspirations from the mature monkeys investigated by Switzer (1967a, Table 8).

Tibial aspiration biopsies were obtained from 10 approximately 1-year-old *Macaca mulatta* that were then placed on a vitamin E–deficient diet for 17–28 months (Porter et al. 1962). In addition to anemia and signs associated with vitamin E deficiency (e.g., muscular dystrophy), marrow biopsies performed at the end of experimentation displayed a mild normoblastic hyperplasia. The basal mean level of normoblasts was 24.8% while the mean postexperiment level attained 38.8% normoblasts, n = 5 (Table 8).

The distribution of erythropoietic activity, as estimated by the localization of intravenously injected tracer ^{59}Fe, was studied in male, young adult commercially imported rhesus monkeys (n = 7, approximately 4 years old, weight 3.5–4.5 kg) (Taketa et al. 1970). The animals were injected with ^{59}Fe-ferrous citrate solution and 25 hours later, they were exsanguinated under barbiturate anesthesia while being infused intravenously with normal saline. After cardiac arrest, the thoracic and abdominal viscera were removed. The bones and bone groups (e.g., vertebrae) were removed and cleaned. The radioactivity of the bone marrow (while enclosed in the bones) was determined by scintillation counting. The ^{59}Fe activity was computed as the percentage of total body osseous-located activity. As a group, the vertebrae were the most active bones, contributing about 33% of the total marrow activity (lumbar vertebrae sequestered 17%), the lower limbs (pair of femurs 13%, paired tibiae 7%), paired upper limbs (12% [right and left humeri 9%]), hip bones (13% [and the relatively inactive sacrococcygeal bones 2%]), skull and mandible (11%), ribs and sternum (6%), and scapulae (4%). Since the injected tracer iron is believed to be preferentially localized in the developing erythroid cells, the distribution of the radioactive iron quantified the relative proportion of total body erythropoiesis located at a given bone or group of bones.

The preceding discussed studies have demonstrated that the hemopoietic bone marrow in the rhesus monkey has much the same distribution as in man. (The current most-utilized site for an aspiration marrow biopsy in the adult human is the iliac crest. The prior most-selected site was the sternum.) The marrow of the sternum, vertebra, and rib of the monkey is "red," that is, actively hemopoietic and red in color as demonstrated by the imprints obtained by Stasney and Higgins (1936, Figure 20). But the marrow of the midshaft of the femur and particularly the tibia is yellow in color ("yellow marrow") due to the presence of a large number of fat cells (Figure 18).

In man in infancy, all bone marrow is actively hemopoietic and is red marrow. It is devoid of fat cells. With age-associated increase in size of the individual, relative expansion of intramedullary hemopoietic space (e.g., increase of length and size of the femur and tibia), and cessation of a progressive need for a larger mass of circulating erythrocytes, fat cells begin to appear in red hemopoietic bone marrow. In addition, the distribution of red marrow contracts and many of the bones discontinue the production of hemic cells and display only fat cells (yellow marrow). Upon the gradual attainment of adult stature, the red bone marrow is limited principally to the flat bones of the skull (calvarium), vertebrae, ilium, sternum, ribs, and possibly proximal regions of the humerus and tibia. In addition, the red marrow, though grossly red in color, nevertheless at this point in life is comprised on a volumetric basis of 50% adipose cells. This adipose space serves as a potential site for increased erythropoiesis (or granulopoiesis) at the time of physiologic need or pathologic conditions (e.g., chronic myelogenous leukemia). The quantity of fat in man's red marrow is used as an indicator of the level of proliferative activity. That is, the less fat the greater is the production of blood cells.

The bone marrow studies of Stasney and Higgins (1936), which as cited previously are devoid of any technical error due to a dilution of the sample with circulating blood cells, show that the bone marrow of the sternum, rib, and a vertebra all demonstrate essentially the same relative proportion of developing erythroid cells (mean 41% of the hemic population) regardless of the fact that they represent samples of different proportions of the body's total erythropoiesis. It is also universally accepted among hematologists that the distribution ratios of hemopoietic bone marrow cells do not depend on the site of red bone marrow (i.e., which bone is biopsied). That is, the same hematocellular quantitative relationships are maintained throughout a monkey's (and man's) hemopoietic bone marrow (Lapin and Cherkovich 1972).

The proportion of adipose cells in the hemopoietic (red) marrow across the age and growth spectrum of the infant, young, and adult macaque has been given sparse specific attention. As indicated previously, yellow marrow has been cited in the tibia and femur (Stasney and Higgins 1936).

The investigation of Suarez and coworkers (1942, 1943) which included infants, young, and adult rhesus monkeys revealed that, in at least their subjects, the youngest monkeys present the highest proportions of normoblasts in the bone marrow (Table 8). And that with the increasing age, the relative proportion of normoblasts in the marrow diminishes. The lowest frequency of normoblasts in the bone marrow smears was obtained in mature adult monkeys. However, it has been noted that the incidence of normoblasts (generic erythroid precursors at all levels of maturation) reported by Suarez for various ages of monkeys was the lowest in comparison with that described by other hematologists (Table 8). This may be due to the technique in obtaining the samples. If the samples were diluted with an undue amount of blood, the sample could contain a large number of circulating leukocytes that would reduce the relative number of normoblasts in a given sample. Suarez and his coinvestigators (1942, 1943) stated that a very strong negative pressure was needed to obtain (aspirate) a marrow sample. This would support the possibility of attendant aspiration of blood into the sample (Cohen 1953). In addition, the work of other investigators suggests that the incidence of erythroid precursors is lower in the bone marrow of young immature rhesus monkeys than in mature adults. This concept appears to be supported by the assembled data of Table 8. The approximate mean value for the representation of normoblasts derived by Winkle (1951) and Porter et al. (1962) for their healthy 1-year-old subjects is ~26%. The intermediate aged 1–2-year-old monkeys evaluated by Usacheva and Raeva (1963) demonstrated ~31% normoblasts in their marrow differential counts. Finally, Switzer (1967a) assessed mature subjects (males mean 5.8 years old, females mean 8.2 years old) and observed that the mean normoblast level was ~40%. Huser's data (1970, Table 8) are likewise in this range. The modest-sized cohort of adults studied by Stasney and Higgins (1936) yielded ~41% normoblasts (sternal, rib, and vertebral samples). The hemopoietic bone marrow of the mature rhesus monkey thus contains a higher proportion of normoblasts than adult human red marrow. This status also seems to be obtained in the immature monkey. The normal human red marrow (depending on the reporting source of information) maintains a mean 18%–26% normoblasts.

The pattern of a higher level of developing erythroid cells in a macaque's hemopoietic bone marrow versus that of man positively coincides with the recognition that the erythrocyte count of the monkey is considerably higher than that of the human, and the fact that the life span of the erythrocyte in the monkey (~100 days) is shorter than the life span of the red cell in man (120 days). One could therefore assume that the monkey's marrow bears the requirement of maintaining a greater relative number of proliferating normoblasts to sustain this bigger burden.

The meager available data suggest that a sex-linked difference in the number of normoblasts in males as opposed to females in the erythropoietic bone marrow of either immature 1-year-old monkeys (Winkle 1951) or adult monkeys (Switzer 1967a) is not statistically identifiable. Thus, although under certain circumstances the circulating erythrocyte count is higher in adult males of this species than in adult females, it can be assumed that this difference is either not reflected in the number of developing erythroid cells (normoblasts) in the marrow in the two sexes or that if a veritable difference does exist, the aspiration biopsy technique followed by differential counts performed on stained dry film smears is an imperfect sampling technique to identify the presence of the sex-linked bias.

The range of the occurrence of normoblasts in normal hemopoietic bone marrow of *Macaca mulatta* is very broad. The range is roughly 10%–50% (Table 8). Although the data are minimal, it appears that the value tends to be higher in adults. In some instances, lower values may be due to a dilution of the aspirated sample with circulating blood. The values of Winkle (1951) would probably not be subject to this error as she specifically indicated that small samples were routinely taken. Nevertheless, the range in her year-old subjects began at 10%–13%. The 31%–48% range reported by Switzer (1967a) for 25 adult *Macaca mulatta* (5 males and 20 females) is noteworthy because of the high "starting" level for the normoblast content and the investigator's statement "careful attention was given to enumerating cells in and around the periphery of bone marrow particles to minimize errors due to dilution with peripheral blood elements." Unfortunately, Stasney and Higgins (1936) did not report the ranges for the incidence of normoblasts in their adult monkeys.

The identification of aspirated marrow samples that have been diluted with a greater than usual amount of peripheral blood is likely to have a subjective component. As noted earlier, the enumeration of hemopoietic cells in the region of a bone marrow particle helps insure that these cells are truly derived from bone marrow stroma. (Aspirated bone marrow typically yields pale white granular macroscopic clusters of cells. These are the so-called bone marrow particles or units. When they are flattened and dispersed in the process of making bone marrow smears, many of their cells are seen surrounding small residual aggregates of cells that are too compact to accurately enumerate in the differential count. Such free nearby cells can be considered likely to have been derived from the unit and consequently are bona fide marrow-residing cells.)

The ideal site for a marrow aspiration biopsy in the adult rhesus monkey seems to be related to an investigator's personal choice. The ischium and ilium appear to yield predictable, verifiable, and representational samplings (Table 8). Switzer (1967b) has discussed the conduction of and his preference for ischial biopsies.

CHAPTER 8

Erythrocyte Life Span

The life span of the erythrocytes of a healthy *Macaca mulatta* is approximately 100 days (Kreier et al. 1970). The latter investigators derived the value of 99.6 ± 1.0 days in a study of four subjects weighing 2–3.5 kg and which, according to their dental development, were 3–6 years of age. The erythrocyte life span was established by labeling a pool of an individual's newly synthesized red cells with the isotopic label $DF^{32}P$ (radioactive phosphorus-labeled diisopropyl fluorophosphate). The label was injected intramuscularly and was subsequently incorporated into the hemoglobin of developing erythrocytes. The dose was based on that used for man, 0.1 mg $DF^{32}P$/kg of body weight. Samples of blood were obtained 1 hour post-$DF^{32}P$ injection and thereafter at 1–3-day intervals for 100 days. The erythrocytes were washed with saline, frozen, lyophilized, digested, dried again, solubilized, and mixed with scintillation fluid. The radioactivity was quantitated by liquid scintillation spectrometry. $DF^{32}P$ is a particularly effective label because its elution from tagged red cells is essentially nil, thereby permitting an accurate determination of the life span of the circulating labeled red cells. The random loss of $DF^{32}P$ label from the circulation was 0.078% per day.

Studies of the survival of circulating autologous chromium-51-labeled red cells in the rhesus monkey conducted along with the independent study of $DF^{32}P$-labeled red cells in the monkeys discussed earlier (Kreier et al. 1970) have led these authors to conclude that the life span of the red cells of *Macaca mulatta* as derived per sodium chromate labeling is 86–105 days. The mean elution rate of the chromium from the tagged cells was 3.15% per day. The half-life of the circulating ^{51}Cr-labeled erythrocytes extrapolated from the published graph was ~17–18 days. The technique for ^{51}Cr labeling of erythrocytes obtained from the circulating blood that was utilized by Kreier et al. included the following steps. A sample of blood was obtained from a given rhesus monkey; the red cells were incubated with radioactive sodium chromate and reinjected intravenously into the donor. The survival of the labeled autologous red cells was monitored for 110 days with the sample taken at 24 hours considered as the 100% level. The radioactivity of the tagged circulating red cells was determined by lysing collected erythrocytes and assaying for ^{51}Cr in the resultant solution of hemoglobin by gamma spectrometry with a well-type, solid scintillation detector.

The mean half-life of ^{51}Cr-tagged autologous erythrocytes in *Macaca mulatta* derived from an investigation of six healthy young male subjects by Glomski et al. (1971) was similarly determined to be 17.1 (s.e. $\pm$ 0.28) days. In the experience of Marvin et al. (1960), one normal "rhesus type" monkey was observed to display a ^{51}Cr-tagged autologous erythrocyte half-life span of approximately 19 days and a maximum survival time of 98 days.

The best-recognized label that can bind to erythrocytes and be used to monitor their survival in the circulation is radioactive sodium chromate (^{51}Cr). Chromium penetrates the red cell membrane and once inside the cell it binds to the hemoglobin. The other isotopes used in establishing a subject's erythrocyte life span conduct their labeling by being incorporated into the hemoglobin molecule as it is synthesized in new cells. Chromium binds to erythrocytes *in vivo* as well as *in vitro*. The anionic (hexavalent) chromium in the form of the chromate ion (^{51}CrO$_4^{-2}$) passes through the red cell membrane and once it is within the cell, it is converted to the trivalent cation (Cr^{+3}) that binds to the globin moiety of hemoglobin. This form of the ion will not be reutilized and become incorporated into other red cells upon release from a labeled cell when the latter dies because it is unable to penetrate the plasmalemma of the red cell. The liberated isotope is excreted (in the urine) thus making some of the calculations more straightforward. When samples of erythrocytes are labeled *in vitro*, ascorbic acid is added to the suspension at the end of the incubation period to convert any surplus chromate that has not been incorporated into the red cells to the reduced trivalent state. This prevents any further labeling of other erythrocytes when the sample is introduced into the animal. A disadvantage of this label is that its affinity for hemoglobin though strong is not irreversible. This so-called elution has been best recognized in human erythrocytes; the usual assumed rate of loss in man is about 1% per day. The half-life of autologous ^{51}Cr-tagged red cells in the marmoset (*Tamarinus nigricollis*) is 15.9 days (n = 14, Merritt and Gengozian 1967). The half-life of ^{51}Cr-tagged-labeled autologous erythrocytes in chimpanzees *Pan troglodytes* (n = 6, mean weight 20 kg), male gibbons *Hylobates lar entelloides* (n = 6, mean weight 3.5 kg), and male baboons *Papio cynocephalus* (n = 6, mean weight 14 kg) were all reported to be 14 days (Rowe and Davis 1972). The zero time and 100% level of radioactivity were established at 0.5–1 hr postinjection of the tagged red cells. Since it is believed that there is an initial higher rate of loss of the label due to its failure to bind or weak binding, or other nonspecific causes during the initial 24 hours the cited red cell half lives in this study may be slightly shorter than the values reported in other investigations (i.e., a slower calculated rate of loss of label is derived if 100% radioactivity is the level obtained at 24 hr postinjection). A high elution rate of chromium appears to be a feature of nonhuman primates, and the degree of elution of this isotope from their labeled erythrocytes has not been assessed in particular detail.

The chromium labeling of erythrocytes does offer one special feature that is typically not afforded by other red cell labels. That is, this element can be used to simultaneously identify two different pools of red cells with labels that are physiologically indistinguishable but can nevertheless be monitored within one test subject for any given duration. This is accomplished by tagging one test pool with the routinely utilized ^{51}Cr-labeled sodium chromate and labeling the other test pool with sodium chromate that has been tagged with the stable, nonradioactive isotope [50]Cr. This latter isotope can be monitored and quantitated following recovery from the subject by neutron activation (neutron bombardment in a nuclear reactor) that converts it to the radioactive isotope ^{51}Cr which is then quantitated by high-resolution gamma spectrometry. In one investigation, the half-lives of Cr-labeled autologous erythrocytes of six healthy male rhesus monkeys *Macaca mulatta* were determined first by [50]Cr-neutron activation analysis and 3 months later using [51]C tagging. The mean T½ of [50]Cr-labeled autologous erythrocytes was 17.8 (s.d. ±1.1) days, a value that is in close agreement with the ^{51}Cr-derived half-life of 17.1 (s.d. ±0.8) days for the same subjects. Neither approach was found to yield consistently higher or lower values. In addition, the data derived from each monkey were subjected to a nonlinear least-squares analysis. In this instance, the mean fitted curves still approached identity with a chromium T½ of 17.1 (s.d. ±1.4) and 17.9 (s.d. ±2.7) days, respectively, for ^{51}Cr and [50]Cr labeling (Glomski et al. 1971b). Figure 21 illustrates the rate of disappearance of [51]Cr- and [50]Cr-tagged erythrocytes from the peripheral blood of these discussed monkeys. The half-lives of the different erythrocytes in combined pools of [50]Cr- and ^{51}Cr-labeled autologous erythrocytes as well as commixtures of ^{51}Cr-tagged autologous and [50]Cr-labeled homologous red cells were compared simultaneously in eight human pediatric patients and one healthy adult male human. Both isotopes yielded equally effective results. In the three patients with sickle cell disease that were transfused with homologous erythrocytes labeled with [50]Cr and autologous cells tagged with ^{51}Cr, the data revealed the capability of this technique to reveal differences in survival between two distinct populations of red cells monitored simultaneously in the same individual (Glomski et al. 1976).

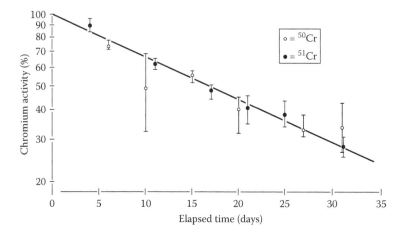

Figure 21 *Macaca mulatta*. A comparison of the rate of disappearance from peripheral blood of [50]Cr- and ^{51}Cr-labeled autologus erythrocytes in six young male rhesus monkeys *Macaca mulatta*. Each point represents the mean value of six pairs of samples. The bars represent the minimal and maximal range of activity. (From Glomski, C.A. et al., *J. Nucl. Med.* 12, 31, 1971.)

Marvin et al. (1960) as well as Fitch et al. (1980) have shown that vitamin E–deficient monkeys present a shortened survival of their ^{51}Cr-labeled erythrocytes that are subsequently lengthened upon recovery from the induced anemia. *Macaca mulatta* that were intraperitoneally injected with a 2.5% aqueous solution of methylcellulose (biweekly over a 14-week period for a total dose of 5.6 g/kg of body weight) developed a mild decrease in their hemoglobin level along with a diminished mean half-life of their ^{51}Cr-labeled autologous erythrocytes. The pretreatment mean half-life span was 17.1 days as opposed to the final mean of 13.3 days (P < 0.025), measured 1 month after final injection (Glomski et al. 1982). These studies help substantiate the utility of ^{51}Cr labeling in the evaluation of the relative life spans of *Macaca mulatta* erythrocytes under experimental conditions.

The life span of the erythrocyte in the rhesus monkey has also been established by the long-term monitoring of the reticulocyte count following a massive blood loss (Harne et al. 1945). Though this approach is impractical for routine analyses, it does validate the red life spans derived from more contemporary sophisticated techniques. Five adult monkeys that were trained to undergo daily reticulocyte counts from the blood of the ear were phlebotomized a volume of their blood equivalent to 1% of body weight and were monitored thereafter by daily reticulocyte counts for approximately 130 days. The resulting data curves were seen to resolve into three phases. The first extends over the initial period of 20–30 days. During this phase, the reticulocytes rapidly increase in number and then subside to the control level or lower. In the second interval, the reticulocyte count remains at approximately control level. Its onset is at the end of the first phase and persists to about the 80th-day post-phlebotomy following which a recrudescence of reticulocytes produces a second elevation of their numbers. This spontaneous elevation and its return to the control level constitute the terminal segment of the response to the blood loss. The first (and anticipated) elevation of the representation of reticulocytes in the circulating blood is interpreted as a mass influx of red cells from the bone marrow to replace the loss of erythrocytes caused by the phlebotomy. The second period is regarded a stabilized phase in the reticulocyte count following the restoration of the blood picture to normal status. The third period represents a spontaneous reticulocytosis, the function of which is to replace the group of cells released during the first period following the hemorrhage and that are disappearing (at the end of their life span) more or less in mass from circulation. The time lapse between the acute hemorrhage and the peak of the spontaneous reticulocyte reaction is interpreted to be the life span of the red blood cell in the monkey. The values derived in this investigation were about 100 days and ranged from 94 to 117 days (Figure 22). The response of individual monkeys varies but analyses of the data do reveal the three-phase reaction.

Insightful workers have recognized that the diverse erythrocyte life spans among inframammalians and mammalians are likely to reveal a generalized positive correlation between a given species' basal heat production (calories/kg of body weight/day) and its red blood cell turnover (i.e., the percentage of the total circulating red cell population that is replaced each day). Rodnan et al.'s (1957) calculations based on their own erythrocytic data as well as those of others for an assortment of species (reptilian, avian, mammalian) did indeed reveal such a statistically substantiated correlation for

Figure 22 *Macaca mulatta*. The reticulocyte response of a *Macaca mulatta* phlebotomized of a volume of blood equal to 1% of its body weight. Each datum point signifies a reticulocyte count (quantitated here as the number of reticulocytes per 1000 erythrocytes). The reticulocyte count is monitored for approximately 20 days prior to phlebotomy and for more than 130 days thereafter. This graph illustrates a prompt, marked reticulocyte response following the loss of blood, a stable slightly elevated reticulocyte level from post-phlebotomy day 20 until day ~108, and a spontaneous elevation of the reticulocyte count during post-phlebotomy days 110–117. The latter peak is considered to be due to the age-related loss of the cohort of red cells that were released from the bone marrow as reticulocytes in response to the blood loss. The derived life span of the erythrocytes in this subject would be considered to be 117 days. The subject is a 6-year-old male in good health. (From Harne, O.G. et al., *J. Lab. Clin. Med.*, 30, 247, 1945.)

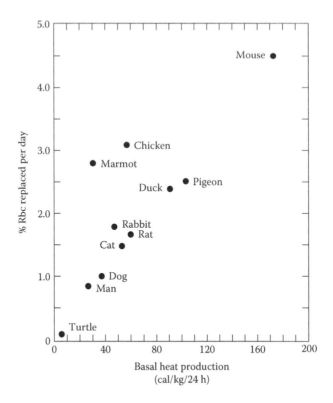

Figure 23   The relationship of erythrocyte turnover (percent of a species' red cells replaced per day) to the basal heat production (cal/kg/day) in a spectrum of species. A positive correlation between the metabolic rate and red cell turnover of subjects was identified as determined by the rank coefficient of correlation P < 0.003. (From Rodnan, G.P. et al., *Blood*, 12, 355, 1947.)

their assemblage (P < 0.003). This relationship is graphically visualized by Rodnan et al. (1957) in Figure 23. Although Rodnan's survey did not include the rhesus monkey, its probable location in the graph would be at the intersection of data points of the y-axis (ordinate): 1.0 (% Rbc replaced per day) and the abscissa (x-axis): 50 (basal heat production calories/kg/24 hr). This proposed locus is based on the rhesus monkey red cell's life span of 100 days and the average metabolic rate of 48.5 cal/kg/24 hr for a mixed-sex population of 11 young rhesus monkeys weighing 2.7–3.6 kg, as determined by Rakieten (1935). This coordinate is consequently located slightly to the right of the tight interval between the sites for the dog and man (Figure 23).

It has also been further recognized that the life span of the erythrocyte is influenced by the body weight of the species. The erythrocytes of small mammals have shorter life spans than those of large mammals. One of the relationships that is functional here is that the smaller the animal, the greater is its body surface area in proportion to its weight. Second, the intensity of the body metabolism has to be proportionally relatively enhanced in smaller subjects to maintain the greater surface area. Erythrocyte

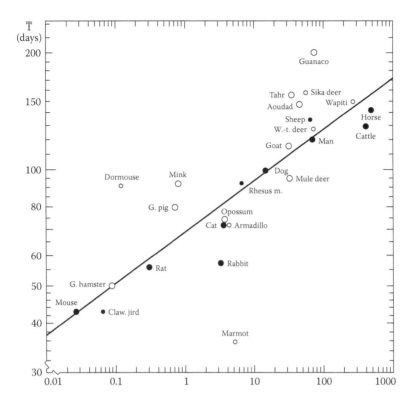

Figure 24 A graphic illustration of the relationship that is identifiable between the mean life span (in days) of mammalian species' erythrocytes and the mean body weight (in kg) of the given species. The datum point for the rhesus monkey's red cells is in the approximate center of the field, and that for man's erythrocytes is nearby. The small white (open) circles correspond to the least reliable red cell life span values (available at the time of the formulation of the graph). The small, black circles designate an increased level of documentation while the large, white circles are accorded to the species' erythrocytes whose life spans have an even greater scientific certainty. The large, dark circles indicate the most reliable values. The dark line is the calculated regression line. The linear approximation is obtained from log–log coordinates of the weights and red cell life spans. T = mean life span of an erythrocyte in days; X axis = body weight in kilograms. (From Vacha, J., Red cell life span, in Agar, N.S. and Board, P.G., eds., *Red Blood Cells of Domestic Mammals*, Chapter 4, Elsevier, Amsterdam, the Netherlands, 1983, Figure 4.9, p. 120.)

life span is prolonged as a result of a decreased rate of metabolism and shortened with the experimental heightening of the intensity of metabolic events (Vacha 1983). The dependence of the red blood cell life span on the mean body weight of the mammal has been investigated by Vacha (1983). His analysis that includes *Macaca mulatta* resulted in identifying a log–log relationship between the length of the mean red cell life span in days and the weight of the species in kilograms (Figure 24). Thus, as illustrated in the latter figure, the erythrocyte life spans of the mouse, rat, rhesus monkey, man, and horse progressively increase as do the weights of the subjects.

CHAPTER 9

Blood Volume

The total blood volume of the rhesus monkey (n = 23) derived by dilution of T-1824 dye and determination of the hematocrit by centrifugation in a Wintrobe tube was calculated to be 75.1 ± 12 mL/kg of the body weight (Overman and Feldman 1947). In another assessment also utilizing Evans blue dye (T-1824) coupled with ^{32}P-labeled-erythrocytes dilution, the ratio of the volume of blood to the unit weight of the rhesus monkey was determined to be a mean 54.0 mL/kg (n = 18, 6–9 months-old subjects) (Gregersen et al. 1959). Growth and increase of body weight monitored for more than 1 year (n = 2) did not alter the ratio. The quantitation of the blood volumes using ^{51}Cr-labeled erythrocytes in one control and four anemic vitamin E–deficient rhesus monkeys yielded values of 55 and 57 mL/kg, respectively (Fitch et al. 1980). The blood volume of the rhesus monkey (n = 10♂ and 10♀, prepubertal subjects, 2.2–5.3 kg) derived by the dilution of radioactive iodinated human serum albumin following a 10-minute postintravenous injection equilibration was determined to be a mean 60.9 mL/kg (Bender 1955). The probable error was estimated to be 1 mL/kg. The value of 61 mL/kg has been utilized or cited by Krise and Wald (1959), Bourne (1975), and Mandell and George (1991). In their investigation of the normal distribution of the cardiac output in the rhesus monkey, Forsyth et al. (1968) calculated the circulating blood volume to be 89.5 mL/kg (n = 6). The concept that the blood volume of the individual is in the range of 6% of one's weight appears to be applicable to both the monkey and man.

The mean plasma volume for the rhesus monkey as established by Gregersen et al. (1959) is 36.4 mL/kg. The earlier study of this primate by Overman and Feldman (1947) yielded an average volume of 44.9 ± 7 mL/kg (n = 23).

The total blood volume of cynomolgus monkeys as determined by the Evans blue dye dilution was found to be 6% of the animal's body weight (Ageyama et al. 2001).

CHAPTER 10

Megaloblasts

Megaloblasts are an abnormal lineage of developing erythroid cells with specific morphologic features that allow their ready identification, typically in Romanowsky-stained (Wright's stain) dry film smears of blood or bone marrow (Figure 25a,b). Megaloblasts are identifiable in the bone marrow of *Macaca mulatta* in certain pathologic conditions and display the identical morphologic alterations seen in human megaloblasts (e.g., as in pernicious anemia, megaloblastic anemia of infancy). Megaloblastic erythroid cells have not been found in the marrow of normal monkeys (Sundberg et al. 1952). The association of megaloblastosis and experimental disease in the rhesus monkey is discussed later in the text.

Historically, the initial recognition of the megaloblast has been attributed to Ehrlich (circa 1880), although one or two individuals recognized abnormal cells in pernicious anemia before his report (Sundberg et al. 1952). Although a multitude of morphologic features characterize the megaloblastic series, the term *megalo blast* in its most straightforward sense refers to the larger size of these cells in comparison with the normoblasts of normal erythrocytic development. The megaloblastic line of development is recognized as a singular, complete line of erythroid development that morphologically begins with the counterpart of the pronormoblast entitled the promegaloblast. The latter cell continues its maturational progression to the end-stage denucleated erythrocyte. The intermediate stages of megaloblastic maturation have been assigned titles comparable to those of the normoblastic series (Table 9). Under frequent conditions, the derived erythrocyte adheres to the *mega* character-ization by being a macrocytic (also often oval) erythrocyte. The erythrocytes derived via megaloblastic differentiation have been termed *megalocytes* by some investigators. Similarly, the term *macrocyte* has been routinely and logically applied to them. Nevertheless, the size of megaloblasts is extremely variable.

The developing erythrocytes of the normal or of the megaloblastic marrow of the rhesus monkey resemble the equivalent cells of man to an inordinate degree (May et al. 1950a; Sundberg et al. 1952). The differences are indistinct, and a morphological analysis of the monkey's megaloblastic differentiation of its red cells serves equally well for man's megaloblasts. Megaloblasts, as cited earlier, go through the same developmental stages as the normoblasts. However, there is a lesser tendency for the megaloblasts to reduce their cell size with maturation than is observed in normoblasts.

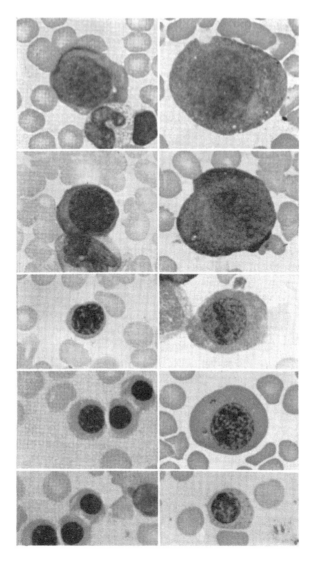

Figure 25a *Macaca mulatta*. (a) A photomicrographic comparison of the morphology of normoblasts and megaloblasts as observed in Wright-stained dry film smears of
bone marrow from normal rhesus monkeys and rhesus monkeys with megaloblastic anemia. The vertical column on the left side of the plate presents the
normoblastic maturation of the red blood cell, while the right column of cells
illustrates megaloblastic development. The youngest, most immature cells, that
is, the pronormoblast or promegaloblast, respectively, are at the top of their columns. The subsequent lower cells illustrate the progressive maturation of the red
cell precursors. The larger size of the megaloblasts and the characteristic more
open, less condensed nuclear chromatin pattern of these cells are readily apparent. Oil immersion microscopic magnification. A black-and-white photograph of
an artist's colored drawings of these very same cells is presented in the adjacent
page (Figure 25b). (From Sundberg, R.D. et al., *Blood*, 7, 1143, 1952.)

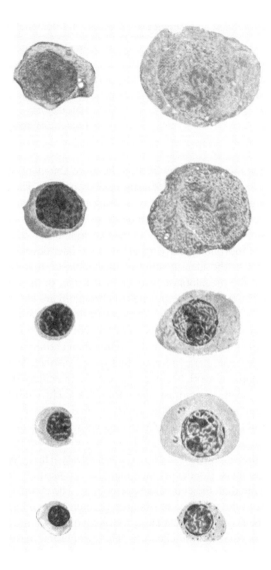

Figure 25b *Macaca mulatta.* (b) This is a black-and-white photograph of a medical artist's (Ms. V. Moore) colored drawing of the very same, identical cells presented on the adjacent page, Figure 25a. This plate thus presents a comparison of the morphology of normoblasts and megaloblasts as observed in Wright-stained dry film smears of bone marrow from normal rhesus monkeys and rhesus monkeys with megaloblastic anemia. The vertical column on the left side of the plate presents the normoblastic maturation of the red blood cell, while the right column of cells illustrates megaloblastic development. The youngest, most immature cells, that is, the pronormoblast or promegaloblast, respectively, are at the top of their columns. The subsequent lower cells illustrate the progressive maturation of the red cell precursors. The larger size of the megaloblasts and the characteristic more open, less condensed nuclear chromatin pattern of these cells are illustrated. Oil immersion microscopic magnification. (From Sundberg, R.D. et al., *Blood*, 7, 1143, 1952.)

Table 9   Normoblast and Megaloblast Maturational Nomenclature

Pronormoblast	Promegaloblast
Basophilic normoblast	Basophilic megaloblast
Polychromatic normoblast	Polychromatic megaloblast
Orthochromatic normoblast	Orthochromatic megaloblast
Reticulocyte	Reticulocyte
Erythrocyte	Erythrocyte (megalocyte)

The cytoplasm of the megaloblasts is likely to be more abundant than it is in normo-blasts. The morphology of the megaloblastic nucleus is often described as generally retaining an immature appearance and presenting other anomalous features during maturation of the cell while the cytoplasm develops normally (as observed in its pro-duction of hemoglobin). The asynchrony in the development of the nucleus and cyto-plasm becomes more morphologically apparent in the later developmental stages and is a principal feature in the identification of megaloblastic maturation. The nuclear chromatin pattern of the megaloblasts is open or net like. In the earliest normoblasts (as in the pronormoblast), minute aggregates of chromatin are identifiable but are less identifiable in the megaloblast. The open network of the chromatin in the megaloblast is associated with easily identifiable masses of parachromatin (amorphous, nonchro-matin areas within the nucleus that stain differently than chromatin) while the normo-blastic counterpart presents parachromatin that is less distinct. With maturation, the megaloblastic chromatin network becomes coarser, the interstices in the net are larger, and the net is more open. In contrast, the normoblastic chromatin pattern becomes more densely clumped and is often more darkly stained. Nucleoli are variable in size, shape, and number in the megaloblast. Consistent with the concept of impaired devel-opment, nucleoli can be observed in more differentiated cells (as in the polychromatic megaloblast) while they are absent in normal normoblasts. The cytoplasm of the mon-key basophilic megaloblast is a deeper blue than the cytoplasm of the promegaloblast and is more basophilic than the equivalent normoblast. The cytoplasmic basophilia of both lineages is due to the presence of ribonucleoprotein, which has the foremost role in the synthesis of hemoglobin. Granular or rodlike white, unstained dots in the blue-stained cytoplasm are mainly mitochondria. The cytoplasmic basophilia of both the monkey basophilic megaloblast and basophilic normoblast tends to be more intense than that of the human versions of the cell. At the polychromatic level of develop-ment, the megaloblast characteristically presents eosinophilic-staining hemoglobin in its cytoplasm along with an open, immature nucleus. This disparity in the maturation of the cytoplasm and nucleus is perhaps the major, most recognizable morphologic fea-ture of megaloblasts. In the final stage of maturation (the orthochromic megaloblast), the cell continues to demonstrate abnormalities. These include karyorrhexis and kary-olysis, multiple Jolly bodies, basophilic stippling, and cytoplasmic inclusions of hemo-siderin. The denucleated erythrocyte is likely to be macrocytic in comparison with the normal monkey erythrocyte and is often oval (as is man's megaloblastic erythrocyte). Table 10 presents a detailed stepwise comparison of megaloblastic and normoblastic differentiation (Sundberg et al. 1952).

Table 10 Comparison of Megaloblasts and Normoblasts of *Macaca mulatta* in Wright-Stained Dry Film Smears

	Megaloblasts	Normoblasts
Size	Extremely variable, but generally larger than normoblasts; inconsistent reduction in size with maturation	Less variable and smaller than megaloblasts; relatively consistent reduction in size with maturation
Nuclear-cytoplasmic ratio	Cytoplasm relatively abundant	Cytoplasm relatively sparse
Nuclear pattern	Vesicular or net like	Vesicular, but even in early forms shows minute aggregates of chromatin
	Open pattern with easily visible parachromatin	Parachromatin distinct but less prominent than in megaloblasts
	Chromatin aggregates small and located at intersections of strands of net	Chromatin may show clumped distribution even in basophilic stage
	With maturation, net becomes more coarse, interstices in net are larger, and net is more open	By polychromatic stage, clumped nuclear chromatin pattern is well established
	Chromatin stains reddish purple, often paler than in normoblast	Chromatin stains rich reddish purple, usually darker than in megaloblast
	Parachromatin is faintly pink	Parachromatin is pink
Nucleoli	Pale blue, variable in size and shape, almost always present in early stages	Pale blue, variable in size and shape, often covered with chromatin
	May persist until late polychromatic stage	Present in most pronormoblasts and in some basophilic normoblasts
Cytoplasm, promegaloblast	Deeper blue than most myeloblasts but paler than basophilic stage	Color comparable to that of megaloblast
	Mottled appearance due to mitochondria	Less mottled
Cytoplasm, basophilc stage	More basophilic than promegaloblast	More basophilic than pronormoblast
	Basophilia sometimes more intense than in normoblasts	Basophilia sometimes less intense than in megaloblasts
	More mottled than promegaloblast	Less mottled
Cytoplasm, polychromatic	Hemoglobin becomes apparent	Hemoglobin becomes apparent
	Color varies with degree of hemoglobinization and of original basophilia	Color varies with degree of hemoglobinization and of original basophilia
	Asynchronism common—hemoglobin in cells with markedly basophilic cytoplasm and extremely immature nuclei	Asynchronism uncommon—hemoglobin usually not seen until nucleus is reasonably mature
	Less mottling, cytoplasm smoother	Almost no mottling, cytoplasm smoother

(Continued)

Table 10 (*Continued*)   Comparison of Megaloblasts and Normoblasts of *Macaca mulatta*
in Wright-Stained Dry Film Smears

	Megaloblasts	Normoblasts
Cytoplasm, orthochromatic	Red-orange color and smooth in appearance	Red-orange color and smooth in appearance
Chromosomes	Long and thin	Shorter and thicker
Mitoses	Multipolar common	Only bipolar normally
Loss of nuclei	Some pyknosis, but karyorrhexis and karyolysis common	Pyknosis usual
	Multiple Jolly bodies	Single Jolly bodies
Cytoplasmic	Basophilic stippling common	Basophilic stippling not seen
Inclusions	Hemosiderin common	Hemosiderin not common

Sources: May, C.D. et al., *Bull. Univ. Minn. Hosp. Minn. Med. Found.*, 21, 208, 1950a; May, C.D. et al., *J. Lab. Clin. Med.*, 36, 963, 1950b; Sundberg, R.D. et al., *Blood*, 7, 1143, 1952.

Megaloblasts demonstrate, as noted, a great deal of variation in cell size as well as in morphology. A given marrow presenting megaloblastic differentiation of its erythrocytes will also concurrently display developing nucleated erythroid cells adhering to the normal normoblastic process of development. Morphologists analyzing such bone marrow preparations attempt to identify, classify, and enumerate all these different developing cells. In addition, cytologists dealing with these preparations have observed that different cases of megaloblastosis in the monkey (as well as in man) further present another population of developing erythroid cells whose morphologic features do not satisfy the criteria for either megaloblasts or normal normoblasts. These developing erythrocytic precursors have been classified by various workers as "intermediate" erythroid precursors, intermediate megaloblasts, intermediate erythroblasts, and macronormoblasts. The term *metamegaloblast* was considered by May but was not employed due to the fact that the scientific literature was already replete with terminology for megaloblasts and megaloblast-like cells (Sundberg et al. 1952). The adjective megaloblastoid is frequently applied to cells that have some of the morphologic characteristics of megaloblasts.

The megaloblastic process does not solely involve the erythroid series but rather is recognized to be a panmyelopathy. As a result, recognizable "megaloblastic" alterations are also identifiable in other myelopoietic cells. The neutrophils of megaloblastic marrows in particular show a marked increase in size, hyperlobulation of the nuclei, giant metamyelocytes, and marked evidence of premature segmentation manifested by contorted and bizarre-shaped nuclei as early as the promyelocyte stage. Frequently, there is some vacuolization of portions of the nucleus, and as a result, the individual nuclear lobes of mature neutrophils may be connected by several long delicate filaments of chromatin. The overall pattern is frequently more striking in the monkey than in human megaloblastic bone marrow. An increase in cell size and nuclear lobulation are seen in eosinophils and basophils but less uniformly than in the neutrophils (May et al. 1951; Sundberg et al. 1952). In some of cases of rhesus megaloblastosis, the megakaryocytes demonstrate more twisting and folding of the nuclei than normal megakaryocytes. Multinucleated megakaryocytes are also more numerous than observed in normal bone marrow. The morphologic alterations of the neutrophils are likely to be more readily identifiable than the earliest erythroid megaloblastic changes. This can be of diagnostic assistance. In the early stages of experimental production of megaloblastic anemia in monkeys, when the marrow presents "intermediate" equivocal megaloblastic erythroid precursors along with the characteristic neutrophils associated with megaloblastosis, continued participation in the experiment results in the development of a frankly megaloblastic marrow (Sundberg et al. 1952).

CHAPTER 11

Megaloblastic Anemia

The megaloblastic development of erythrocytes is an abnormal process and results in abnormal and insufficient production of red cells, decreased hematocrit, and a lowered level of hemoglobin, thereby producing anemia (and in some instances, a reduced number of granulocytes and platelets in the circulating blood) (Figure 25a,b). Wills and Bilimoria (1932) and Wills and Stewart (1935) were the first to report that a megaloblastic bone marrow could be experimentally produced in macaque monkeys on the basis of deficient diets. Subsequent extensive rigorously controlled investigations employing *Macaca mulatta* were conducted demonstrating that a vitamin C–deficient diet was capable of routinely inducing a megaloblastic anemia (without the use of drugs or vitamin antagonists) (May et al. 1950a, 1951; Sundberg et al. 1952). Immature rhesus monkeys were placed on *ad libitum* milk diets that were deficient in ascorbic acid (vitamin C) while control monkeys were placed on the same diet plus an additional 50 mg of vitamin C per day. The latter subjects thrived and none developed megaloblastic anemia during control periods of 5 months to 1.5 years. On the other hand, megaloblastic anemia developed in all the monkeys if vitamin C was not given.

The monkeys on the ascorbic acid–deficient diet presented a uniform sequence of symptoms in the development of the megaloblastic anemia. The first 2 or 3 months on the experimental diet the subjects would gain weight, behave normally, and reveal no anemia or tendency for a megaloblastic bone marrow. The first indications of ill health were a gradual loss of weight and a decrease of appetite. Changes in the bones of the wrist characteristic of scurvy (the classic disease of vitamin C deficiency) were found at this time but obvious signs of scurvy such as periorbital and gingival hemorrhages and swelling of the extremities did not usually appear until several weeks later. The hemoglobin level and marrow sometimes remained normal for about 1 week after the presentation of obvious scurvy. Once scurvy became manifest, a series of dramatic changes began. Anorexia became profound, the animal lost all desire to move about, the fur lost its luster, and the gingiva became fetid and necrotic. Leukopenia was common. These and other physical manifestations progressed to the point that some animals required parenteral fluids to prolong their life. In this terminal phase, anemia developed rapidly and the marrow would change from a normoblastic to a megaloblastic cytology. At this point if specific therapy were administered, all of these symptoms and abnormal hemocellular findings were promptly relieved and a rapid return to good health occurred even though the experimental milk diet was not changed. Both sexes were identically susceptible to the induction of the megaloblastic process.

A megaloblastic marrow appeared on average after 103 days on the vitamin C–deficient diet (range 90–130 days, n = 7). The megaloblastic marrow appeared a mean 20 days following the appearance of scurvy (range 13–31 days, n = 7). In one additional atypical case, a megaloblastic marrow appeared 63 days after the onset of the vitamin C–deficient diet while scurvy took 74 days to develop (May et al. 1951).

Studies were conducted to determine the effects of pteroylglutamic acid (folic acid), vitamin C, and vitamin B_{12} on the development of megaloblastic anemia while the monkeys were on the experimental milk diet that was routinely successful in producing this anemia (May et al. 1949, 1950a,b, 1951; Sundberg et al. 1952). These factors were tested alone and in combination. Some of the conclusions of these studies were as follows: (1) megaloblastic anemia is produced in rhesus monkeys when they are placed on a milk diet deficient in ascorbic acid and low in folic acid, (2) folic acid given to a monkey with a megaloblastic bone marrow induced by the milk diet deficient in ascorbic acid causes the marrow to promptly reverse to a normoblastic marrow. If vitamin C is also given to such a subject, the monkey survives and hemoglobin and erythrocyte count return to normal, (3) ascorbic acid administered to monkeys on the ascorbic acid–deficient diet containing the minimal amount of folic acid present in the experimental diet caused the reversion of the marrow from megaloblastic to normoblastic cytology, (4) per administered dosage, folinic acid proved to be more effective than folic acid in the conversion of the megaloblastic marrow to normoblastic marrow, (5) vitamin B_{12} is not effective as a therapeutic or prophylactic agent in regard to experimentally induced (vitamin C–deficient milk diet) megaloblastosis, and (6) the requirement for folic acid is increased in scurvy. The immature scorbutic monkey cannot meet this increased requirement from the supply of folic acid obtainable from the milk diet.

It is unclear whether vitamin C has a direct role in hematopoiesis or if the anemia observed in subjects with vitamin C deficiency is the result of the interactions of ascorbic acid with folic acid and iron metabolism (Wixson and Griffith 1986). It is stated that ascorbic acid is required for the maintenance of folic acid reductase in its inactive form. Impaired folic acid reductase activity results in an inability to form tetrahydrofolic acid, the metabolically active form of folic acid. Failure to synthesize tetrahydrofolic acid eventually leads to the development of megaloblastic anemia. Ascorbic acid will induce a remission of the megaloblastic anemia caused by the milk vitamin C–deficient diet only if sufficient folic acid is present to interact with the ascorbic acid. Folic acid supplementation, on the other hand, protects against megaloblastosis of the marrow but its effect does not prevent death from scurvy. The hematologic changes occurring in folate deficiency can be attributed to the fact that folic acid derivatives play an important part in the biosynthesis of purine and pyrimidine nucleotides (Wixson and Griffith 1986). There is a decreased synthesis of deoxyribonucleic acid precursors in folate deficiency that results in a deficient production of various blood cells.

The morphologic picture of the blood and bone marrow of the megaloblastic anemic monkeys on the experimental milk diet was consistent with that of humans with megaloblastic anemia (May et al. 1950a, 1951; Sundberg et al. 1952). An anemia was present with reductions in the erythrocyte count, hemoglobin level, and hematocrit. Anisocytosis, with some macrocytosis and ovalocytosis, was demonstrable in the majority of the animals. Poikilocytosis was prominent and some but not all monkeys showed the lack of polychromasia seen in most human megaloblastic anemias. Karyorrhexis, karyolysis, multiple Jolly bodies, and basophilic stippling were observed. The erythrocytes were normochromic and appeared to be normally full of hemoglobin. The latter observation is consistent with the concept that hemoglobin production in megaloblastic anemia is normal and the major dysplasia is concerned with nuclear factors (the nuclear-cytoplasmic asynchrony). Nucleated erythrocytes were often seen in the blood and were frequently megaloblasts. Leukopenia was the rule. The enlarged hyperlobulated "pernicious anemia neutrophils" that are a classic presence in human megaloblastic anemia were found in the peripheral blood as well as in the marrow. A feature in developing and mature neutrophils with megaloblastic anemia that is uncommon in equivalent cells of man is the presence of Dohle bodies (Sundberg et al. 1952). These cytoplasmic slightly basophilic structures are now recognized to be light microscopic visible aggregates, which when examined under electron microscopy prove to be small collections of rough endoplasmic reticulum. Occasionally, a moderate decrease in platelets was encountered.

The bone marrow of megaloblastic monkeys is typically hyperplastic. Bone marrows presenting 40%–50% megaloblasts were seen in many animals. Normoblasts and intermediate cells are also likely to be present. Their proportionate numbers vary with the state of development or remission of the disease. The typical "megaloblastic" alterations in the morphology of the granulocytes and megakaryocytes, as described previously, accompany the erythroid picture. Hepatic and splenic imprints of autopsied monkeys with fully developed megaloblastic anemia derived from the experimental vitamin C–deficient diet have demonstrated megaloblasts (Sundberg et al. 1952).

Megaloblastic alterations of the developing marrow erythrocytes are identifiable in anemic young cynomolgus monkeys that have been maintained of a vitamin E–deficient diet containing 8% stripped safflower oil (22% of the calories) for a period of 2 years (Ausman and Hayes 1974). The changes are obtained in the terminal stages of the experimental disease.

CHAPTER 12

Erythrocyte-Based Values and Gender Relationships

A biological truism holds that the male of a vertebrate species is likely to maintain a higher erythrocyte count, hemoglobin level, and hematocrit than the female. This is indeed the documented case in the adult human (Kjeldsberg et al. 1989). This semidictum is also evident in the specific case of a member of the antecedent phylogenetic class (the Avians), the domestic chicken (*Gallus domesticus*), the most numerous of all the birds in the world (reviewed in detail in Glomski and Pica [2011]). Another example is the small male Japanese quail (*Coturnix japonica*) that exhibits a significantly higher Rbc count, Hb, and Hct than the considerably larger-sized corresponding female (Nirmalan et al. 1971; Glomski and Pica 2011).

In an overview of the rather extensive collection of data derived from contemporary analyses of the blood picture of the *Macaca mulatta* utilizing modern electronic techniques (presented in Table 1), the rhesus monkey tends to display major erythroid values (Rbc, Hb, Hct) that are higher in the male in comparison with the equivalent nongravid female. The differences in the values of these parameters, however, are not large and when assessed in most investigations fail to attain validated statistical significance. Conversely, the erythroid values of the female rhesus monkey very rarely suggest that their values are higher than those of the male. Thus, the question at hand is whether a natural bias is found in the male's values, and if so, when and under which circumstances.

The simplest and most straightforward factor that positively correlates with the higher erythroid-derived quanta in the male rhesus monkey seems to be the testosterone level (accompanied with the secondarily associated greater muscle mass of the male). A stimulatory effect of androgen upon erythropoiesis is readily identifiable while it is absent with estrogen. In regard to muscle mass and its correlation with erythropoiesis, a case in point is obtained in the Japanese quail. The mature female Japanese quail is larger and weighs ~20% more than the male counterpart, while the male, nevertheless, has a muscle mass that is ~18% greater than that observed in the female (Wilson et al. 1961). And as noted earlier, the male quail presents significantly higher levels of the primary erythroid parameters (Rbc, Hb, Hct).

An examination of the erythroid data from various studies of the rhesus monkey listed in Table 1 presents only one investigation in which adult male rhesus monkeys generated statistically significant higher values than adult females for the erythrocyte count, hemoglobin concentration, and hematocrit (>3 years old, n = 43♂, 40♀) (Table 1, Stanley et al. 1968[5a,5c]).

In three other studies, two of the three primary erythroid parameters were statistically significantly higher in males. In each case where the item that did not attain a significant difference, the mean numerical value for the male was nevertheless at least minimally higher than that obtained for the female. The ages of the subjects were listed as young adults in one instance, and otherwise from 3–4 years up to 5–10 years old. One hundred fifty-five males and 145 females were assessed in this survey of three independent investigations (Table 1, Robinson and Ziegler 1968[3]; Buchl and Howard 1997[32]; Chen et al. 2009[33]). In Buchl and Howard's study (1997), groups of sequentially aged 3–4-, 4–5-, and 5–10-year-old males and nongravid females were analyzed and in each instance the subsets conformed to the criteria listed earlier.

Five other individual investigations of adult male and nongravid female rhesus monkeys (cited in the following) whose erythrograms did not indicate a verified significant difference between the sexes for any major erythroid parameter presented mean higher levels in males for the Rbc counts in three cases, mean higher hematocrits in four instances, and greater hemoglobin levels in four of the studies. In one set of testings, male bias was not identified in any erythroid descriptor while in three of the investigations, all three red cell–based quantifications were higher in males. These studies involved an aggregate of 54 males and 87 females (Table 1, Switzer 1967a[4]; Lewis 1977[48]; Matsumoto et al. 1980[8]; Kessler et al. 1983[25]; Andrade et al. 2004[1]).

Comparisons of erythrograms of males and pregnant females of similar ages within a given investigation appear to yield some support for the premise that numerically higher values for the Rbc count, Hb, and Hct can at times also be found in the males under this circumstance. The differences, however, are frequently minimal. In those instances where the females offer a greater level for a given parameter the females' increment is usually minor. Free-ranging, living naturally, food-provisioned rhesus monkeys on the island Cayo Santiago, Puerto Rico were studied by Kessler and Rawlins (1983[30], Table 1). The females were typically in midpregnancy as a consequence of the season when they were collected and the natural breeding period of the region. For the cohort composed of members 4–9 years old (36♂, mean 6.2 yr, 15♀, mean 5.8 yr), the males offered higher mean values for the Rbc count and hematocrit. For the group whose members were 10 years old or older (10♂, mean 16.2 yr, 7♀, mean 10.8 yr), the males presented higher average levels for all three major erythroid parameters. In the experience of Buchl and her coinvestigator (1997), the hemograms of mature 3–4-year-old rhesus monkeys (n = 30♂ and 11 pregnant ♀) revealed that the males presented a mean higher red cell count while the females had a greater average hematocrit and hemoglobin concentration (Table 1, superscript 32). With subjects 4–5 years old (n = 44♂ and 20 pregnant ♀), the females presented mean higher values for all three major parameters. However, in the 5–10-year-old age range (n = 21♂ and 44 pregnant ♀) the sex-associated bias favored the males for all three factors.

The sex-linked bias in the red cell–related parameters is apparently associated with the age of the subject. It seems that the higher values in the male become identifiable at maturity or perhaps during the juvenile period (i.e., entry into maturity). This would be consistent with testosterone production. The investigation of Buchl and Howard (1997) is helpful in analyzing the circumstances under which a sex-linked bias may be observed in the erythrogram of *Macaca mulatta* because this study involved a large population of healthy, domestically bred and reared subjects whose hemogramic data were grouped according to age and sex (n = 527). Their compilations showed that the hematocrits of <1-, 1–2-, and 2–3-year-old subjects expressed equivalent values between the sexes during these intervals. For the groups aged 3–4, 4–5, and 5–10 years, the hematocrits of males definitely exceeded those of females. The hemoglobin level was consistently higher in males from 2–3 to 5–10 years of age. The increment between the two sexes was very stable, favoring the males by 0.4 g/dL. During this latter interval, the red cell count of males was persistently slightly and nonsignificantly greater than that of the females.

Since the age at which an individual attains maturity is somewhat variable, it would be conjectured that some juveniles in the 2–3-year-old interval would present evidence of this progression. This could theoretically be indicated by the inclusion of some "adult" male values in the mean hemic profile of a cohort of 2–3-year-old subjects. A set of monkeys that could be included in this conceptual sphere is one of the groupings of Kessler and Rawlins (1983, Table 1, superscript 30). It is comprised of young free-ranging monkeys (18♂, mean age 2.3 yr; 11♀, mean age 2.5 yr). The females are not pregnant and therefore bona fide immature since they were living freely on the island Cayo Santiago and the sexually mature monkeys of such populations are likely to be pregnant. Thus, it is noteworthy that the mean erythrogram of the males of this group displays a significantly higher mean red cell count and higher (but not statistically significant) hematocrit and hemoglobin level than the females. This presentation could be interpreted as potentially due to the fact that some of the males of the group, though not mature, are beginning to approach sexual maturity and are consequently producing some androgens, an activity that is reflected in the male bias of the erythrogram.

Hematologic analysis (complete blood counts) determined by automated instrumentation on umbilical cord blood obtained from newborn rhesus monkeys at cesarean section performed 10 days prior to predicted labor revealed no significant differences with respect to infants' gender (n = 10♂, 7♀, Table 1, Rogers et al. 2005). (The mean length of gestation for *Macaca mulatta* is 164–165 days [Martin et al. 1973; Rogers et al. 2005].)

Winkle (1951, Table 1) reported the hemograms of 20 male and 13 female laboratory-housed immature *Macaca mulatta* that were 10–15 months old. The mean hemocytometer (manual counting chamber)-derived erythrocyte counts were not statistically different in the sexes (♂6.33 × 10^6/μL, s.d., 0.85; ♀6.07 × 10^6/μL, s.d. 0.61, P < 0.3) (established by independent analysis of reported data). The hematocrits and hemoglobin concentrations were identical.

In the extensive investigation of Xie et al. (2013[67]) involving more than 900 nonanesthetized *Macaca fascicularis* (374♂ and 543♀) with an age distribution of 13–72 months (1–6 years), the red blood cell count, hemoglobin, and hematocrit showed significantly higher values (P < 0.01) in males than in females in the

49–60 months and 61–72 months age groups (Table 1, Figure 26). The males in this species reach sexual maturity at approximately 3–4 years of age while females attain sexual maturity at about 3–3½ years (Fortman et al. 2002). Sugimoto et al. (1986a) have reported that "according to the statistics of their breeding colony, the female cynomolgus begins menstruation at 2.5 years ± 8 months on average." Thus, it is seen that the sex-linked bias in the levels of circulating erythrocytes, hematocrit, and hemoglobin concentration of the blood in this species has its onset roughly at the time of sexual maturation. Graphic illustrations of the levels of the Rbc count, Hct, and Hb from the end of the first year through the sixth year of life clearly reveal the onset of a decline of each of these values in the female during the interval of 37–48 months of life that attains statistical significance at 40–60 months (Figure 26). The decline continues and remains significant at least to 72 months (the termination of Xie et al.'s investigation). In addition, the discrepancy between the males' and females' values is further mathematically increased because the values for the major erythroid param-eters in the males start to increase (at least in Xie's colony) at about 37 months and continue to augment to 72 months. The increment is minimal in the red cell count while it is modest but definite in the hemoglobin and hematocrit (Figure 26).

A survey of the individual investigations of the blood picture of mature *Macaca fascicularis* (performed without the use of sedation) of animals 5–6 years old (Matsumoto et al. 1980[8]), as well as younger animals 2.5–3.5 years old (Wang et al. 2012[65]), and those 3–5 years of age (Matsuzawa and Nagai 1994[13]; Kim et al. 2005b[12]; Zeng et al. 2011[49]), all consistently presented slightly higher values for the Rbc count, Hct, and Hb in males (Table 1). Table 5 presents sex-grouped, weighted mean values of these parameters and they illustrate this quantitative erythrocellular bias.

Erythrograms performed on 27 male cynomolgus monkeys with a mean weight of 3.87 kg and 15 females of this species (mean weight 3.14 kg) displayed statistically higher levels of the erythrocyte count, hematocrit, and hemoglobin concentration in the males (Table 1, Schuurman and Smith 2005). Whether or not a sedative was employed to assist in the venisections was not specified. Using the tables of weights and age for laboratory-bred cynomolgus monkeys (Terao 2005), it is assumed that the monkeys were mature (males with a weight of 3.87 kg would be ~4.7 years old while females weighing 3.14 kg would be ~5.6 years old).

A comparison of the results derived from two large populations of laboratory-maintained cynomolgus monkeys (Yoshida et al. 1986a, n = ~1000, age range 1 ≥ 10 yr, Table 1, Xie et al. 2013, n = 917, age range 1–6 yr, Table 1) that were generated approximately 25 years apart offers considerable support for at least one or two characterizations of the erythrocytes in this species. It is noted that the comparison of these two large populations of animals may be imperfect in that Xie's cohort was not administered ketamine while Yoshida's group may have been administered the agent. Both cohorts illustrated that the erythrocyte count was consistently higher in males than in females during the 3–6 years of age interval. However, during this specific period, Xie's male monkeys maintained a stable red cell count while Yoshida's males revealed a slight statistically significant decline in this value. Both cohorts of females demonstrated a decline in the erythrocyte count at this time (Figures 26 and 27). The hemoglobin concentrations statistically

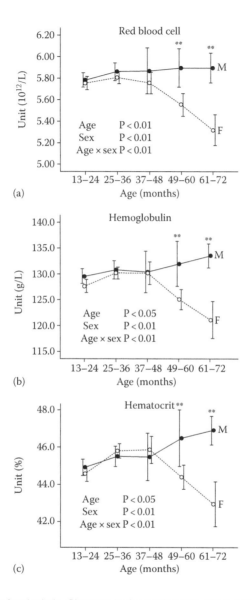

Figure 26 *Macaca fascicularis*. Changes in the erythrocyte count, hematocrit, and hemo-globin in male and female *Macaca fascicularis* with age. The span illustrated begins at 1 year (13 months) and continues through 6 years (72 months). Data are presented as means ± SD at each point. Asterisks indicate a significant differ-ence. M signifies males, F signifies females. (From Xie, L. et al., *PLOS One*, 8(6), e64892, 2013)

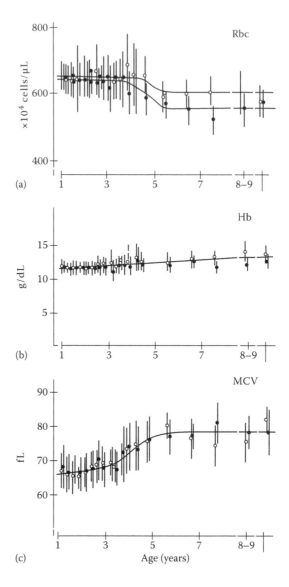

Figure 27 *Macaca fascicularis.* (a) An illustration of the erythrocyte counts of both sexes of *Macaca fascicularis* during the interval from 1–10 years of age (mean ± 1 SD). It is seen that the males (white circles) tended to maintain higher erythrocyte counts than the females (black circles), initiating at ~3.5 years of age. The red cell count diminished significantly in both sexes at ~4.5–5.5 years of age. The recession was greater in females. (b) An illustration of the hemoglobin levels of both sexes of *Macaca fascicularis* during the period of 1–10 years of age (mean ± 1 SD). Males, white circles; females, black circles. The increase with age is statistically significant. (c) An illustration of the MCV of both sexes of *Macaca fascicularis* during the span of 1–10 years of age (mean ± 1 SD). A significant increase is observed in male and females during the ages of 3–5.5 years. (From Yoshida, T. et al., *Exp. Anim.*, 35, 329, 1986a.)

increased with time in both sexes in Yoshida's experiments (age range 1–10 years) but decreased in females in Xie's group. Both Xie et al.'s (2013) and Yoshida et al.'s (1986a) studies revealed a tendency for the mean cell volume (MCV) to increase in the two sexes during the age span of 2–5 years (Table 1, Figure 27). This attained mathematically verified significance in Yoshida's cohort.

An analysis of 274 male and 386 female *Macaca fascicularis* monkeys while they were conscious by Wolford et al. (1986, Table 1) has generated some data that support the occurrence of a male bias in the erythroid blood picture. Although the information was sex identified, the ages of these individually housed laboratory subjects could not be applied to the results because the monkeys had been imported in the wild state and consequently their ages could not be precisely determined. Nevertheless, this population of monkeys of assorted ages presented mean values for the Rbc count, Hct, and Hb that were always higher in the males (not statistically analyzed) (Table 1). This group of cynomolgus monkeys of random ages thus had an erythrocellular profile that was in parallel with most other investigators' findings for mature monkeys. The data reported by the Charles River Laboratories for their cynomolgus monkeys revealed higher sex-linked Rbc, Hct, and Hb levels in males. The number of subjects, their ages, and other potentially relevant information were not indicated in the citation (Table 1, Bernacky et al. 2002).

A summary of hemogramic data collated from 67 member companies of the Japan Pharmaceutical Manufacturers Association covering a population of ~540 male and 500 female wild-caught cynomolgus monkeys between 2 and 5 years old is in agreement with some of the conclusions that were made concerning this primate's blood picture (Matsuzawa et al. 1993, Table 1). The observations, however, are not specific because the analytic techniques are variable. All erythrocyte counts are considered to be electronically based, the microhematocrit technique was used at a few facilities, and the hemoglobin concentration was determined most often as cyanmethemoglobin but also as oxyhemoglobin at other institutions. There was no indication whether ketamine or other medication was utilized to anesthetize the subjects. The monkeys were classified as young adult to adult. Nevertheless, the mean erythrocyte levels were higher in the males ($\male 6.48 \times 10^6/\mu L$, $\female 6.19 \times 10^6/\mu L$) as well as the hematocrit ($\male 41.5\%$ versus $\female 40.1\%$) and hemoglobin ($\male 12.1$ and $\female 11.5$ g/dL) (Table 1). It was specifically noted that the Hct showed greater values in males at 4 years and upward. These data also tend to underscore that the erythrocyte count is higher in the cynomolgus monkey than in the rhesus monkey.

In an independent methodologically modern investigation of 95 male and 95 female 3–7-year-old laboratory-bred *Macaca fascicularis* that were sampled on several occasions, the major erythroid values were very similar in the males and females but the levels were still minimally higher in males ($\male$Rbc $5.49 \times 10^6/\mu L$, Hct 42.9%, Hb 13.2 g/dL; $\female$Rbc $5.44 \times 10^6/\mu L$, Hct 42.2%, Hb 12.9 g/dL) (Table 1, Koga et al. 2005). There was no indication that sedation was employed.

Investigations involving specifically identified ketamine-treated long-tailed macaques (crab-eating macaques) similarly revealed slightly greater, weighted mean age–sex grouped erythroid values in males (Rbc, Hct, Hb). Subjects 3–5 years of age (Sugimoto et al. 1986[21]; Yoshida et al. 1986[66]; Kim et al. 2005a[53]; Kim et al. 2005b[12]) and those 5–10 years old (Sugimoto et al. 1986[21]; Yoshida et al. 1989[9]; Perretta et al. 1991[28]) conformed to this gender-linked format (Tables 1 and 6). The weighted mean values of these parameters for other studied monkeys are also listed in Table 6. For the members of the 8–18-year-old cohort (n = 41♂, 17♀), only the erythrocyte count was higher in the males. Sugimoto et al.'s (1986[21]) conclusions regarding their hematologic assessment of 206 1–18-year-old subjects born and reared at Tsukuba Primate Center for Medical Science, NIH, Japan (plus 32 monkeys of wild origin estimated to be 5 or more years old) affirmed that infant and juvenile monkeys do not present any significant differences in Rbc, Hct, and Hb between males and females (Table 1). However, they also further reported that "these values become significantly larger in males than in females after sexual maturation." In yet another assessment of ketamine-treated cynomolgus monkeys that resided in a mixed-sex caged indoor–outdoor enclosure with inhabitants' ages ranging from 2 months to 15 years old (n = 35♂, 31♀), significantly higher values were yielded for males for all major erythroid parameters (Giulietti et al. 1991, Table 1). A 3-year long survey of healthy adult, laboratory-maintained, ostensibly ketamine-anesthetized *Macaca fascicularis* (n = 89♂, 53♀) presented higher Hct and Hb values in males (Table 1, Adams et al. 2014). Erythrocyte counts were not reported.

It is interesting to note that male sex–linked bias in the erythrogram was minimally identifiable in wild free-ranging *Macaca fascicularis* (Table 1; Takenaka 1981, 1986). The monkeys were trapped, anesthetized with ketamine, examined, phlebotomized within 2 hours of capture, and then released. Young adults and adults of six colonies on Bali Island and Sumatra, Indonesia were evaluated (n = 125). In all six groups, the males of the colony exhibited higher mean Rbc counts, hematocrits, and hemoglobin concentrations. The sole exception was the lower mean erythrocyte count in the males of the Sangeh group (n = 19♂, 11♀).

CHAPTER 13

Erythrocyte Indices (MCV, MCH, MCHC)

The erythrocyte indices are red cell–based mean values that are mathematically derived from the erythrocyte count, hematocrit, and hemoglobin concentration of a blood sample. Their formulas were invented by Wintrobe (1933), and the indices have persisted as elements of hematology since that time. The implications of the indices, however, have been modified since their origin. Wintrobe's hemograms were based on visual counting of erythrocytes in Neubauer counting chambers, and hematocrits were determined by centrifugation. Since then, the electronic counting of the erythrocytes has markedly increased the accuracy of the counts. And centrifugation of blood for the determination of the hematocrit is now recognized to yield imperfectly higher values because of the plasma trapped among the packed cells. Electronic methods of establishing the hematocrit obviate this inaccuracy because the volume of individual red cells is directly measured by the instrument and the hematocrit is mathematically computed from the erythrocyte count and the size of the red cells. It is noted, nevertheless, that under many circumstances, particularly in comparative hematology, the centrifuged microhematocrit is still employed and yields useful clinical information. The spun hematocrit, however, has limitations and is not ideal for many investigative studies.

The erythrocyte indices are readily obtained and easily calculated. They are routinely employed in the characterization of a given individual's or a group's erythrocytes (e.g., members of a genus, males versus females, age cohorts), the delineation of different populations of red cells presented by a given species, and the identification of cytologic alterations that accompany certain physiologic or pathologic conditions. The fact that the indices are derivable by electronic methods (e.g., Coulter counter) as a cost-free, immediately obtainable extension of the erythrogram (Rbc count, Hb, and Hct) has further promoted their universal utilization.

The mean cellular volume (MCV), the average volume of an individual red cell, is expressed as μm^3 or femtoliters (fL). It has proved to be the most clinically and experimentally useful of the three erythrocytic indices. It presents a more accurate designation of the average size of a subject's erythrocytes than obtained from the visual microscopic measurements of red cells in a dry film smear (mean cellular diameter).

The erythrocyte of *Macaca mulatta* as previously noted has a cellular volume in the range of 75 fL that is considerably smaller than the human ~90 fL. However, like man's red cell it is a biconcave disc as verified by scanning electron microscopy (Figure 28) and morphology in dry film smears of blood and bone marrow. An evolutionary reduction of the size of the red cell is usually interpreted as an adaptation that allows it to be a more efficient exchanger of oxygen by virtue of the resultant increase in cell surface per unit of cell volume. The smaller red cells of the monkey can therefore be considered consistent with the fact that the monkey maintains, due to its smaller size and consequent relatively greater body surface area, a higher metabolic rate and accordingly benefits from having "more efficient" gas-exchanging erythrocytes. In fact, a readily observable reduction in red cell size is identifiable in the erythrocytes as they evolve onward from the primitive phylogenetic class Chondrichthyes (cartilaginous fishes) and in particular from the amphibians, to the reptiles, avians, and finally to the mammals (Glomski and Pica 2006, 2011). This phenomenon was first formally illustrated in Gulliver's (1875) classic illustration that surveyed the comparative sizes and configurations of the erythrocytes of the vertebrates (Figure 29). Gulliver (1845) also has the distinction of being the first to describe the monkey's erythrocytes observing that they were smaller than those of man. The optimization of red cell size for a given species and its normal activities is exemplified in the erythrocytes of the dolphin. This aquatic mammal has anucleate, biconcave discoid erythrocytes that are larger than those of man (Ridgway 1972). Although conforming in shape to those found in humans, they are considered less proficient in speedily exchanging oxygen due to their larger size (and consequent relatively less surface area). Here, it is postulated that this geometry permits the

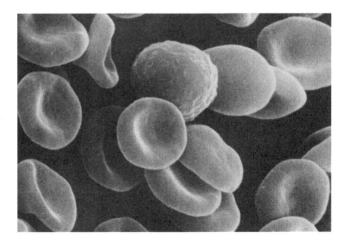

Figure 28 *Macaca mulatta*. Scanning electron micrograph of the erythrocytes of *Macaca mulatta*. They are biconcave discs and consequently have the same morphology as the red cells of man. However, they are ~10% smaller than human red cells. The large spherical cell in the center of the field is a lymphocyte. Magnification ×5420. (From Lewis, H., *Comp. Biochem. Physiol.*, 56A, 379, 1977. With permission from Elsevier.)

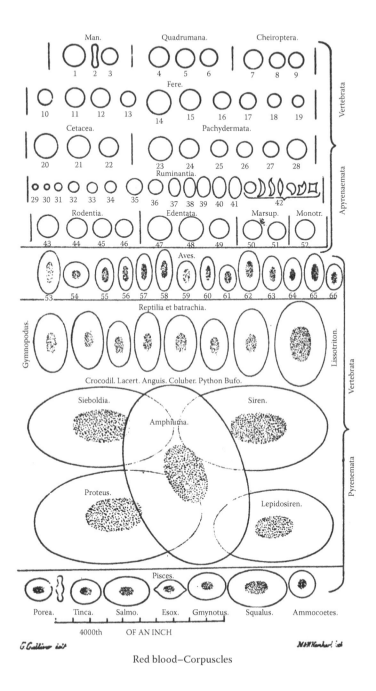

Red blood–Corpuscles

Figure 29 (Continued)

Figure 29 (Continued) The classical landmark illustration by Gulliver (1875) that first surveyed the comparative sizes and configurations of erythrocytes in vertebrates. The variation in the size of the erythrocytes observed among various taxonomic Classes of animals is strikingly obvious. The permanent retention of the nucleus in the erythrocyte by all vertebrates phylogenetically prior to and including the avians (Aves), as well as the oval configuration of their red cells was recognized and illiustrated by Gulliver. The progressive decrease in size of the erythrocytes that initiated with the huge red cells of the primitive urodeles (amphibians, e.g., the *Amphiuma*, illustrated in the lower central field) and continues roughly up to the appearance of the mammalians is also documented. The latter amphibian's red cells are noteworthy in that its RBCs are large enough to be visible to the naked eye. The denucleated status of mature mammalian erythrocytes is also identified. (The mammlians are classified here as Vertebratra Apyrenaemata.) The first six red cells in the top row of the figure depict man's erythrocytes and those of three nonhuman primates. The first cell in the uppermost left is a surface view of the human red cell; the second figure is a human erythrocyte viewed "on edge"; while the third figure is a cell after extraction of its hemoglobin. The next three centrally located cells of the top row (numbered 4, 5, and 6) are nonhuman primate red cells. They are classified as members of the Quadrumana, an obsolete classification that included the apes and monkeys recognizing them as a species with "four hands" because their hind feet are usually prehensile in contrast to man who is bimanual (and has 2 feet). The fourth cell is that of *Simia troglodytes* (*Pan troglodytes*, the common chimpanzee); the fifth is obtained from *Ateles ater* (a spider monkey, perhaps *Ateles fusciceps*, the black-headed/faced spider monkey); while the sixth is a red cell from *Lemur anguanensis* (a lemur whose species title has been discontinued). The erythrocytes illustrated in this figure were drawn to scale. The erythrocytes were studied in suspension and/or dry films. Each small division in the scale at the base of the drawing represents four-thousandths of an inch. (From Gulliver, G., *Proc. Zool. Soc. Lond.*, XXXI, 474, 1875.)

dolphin's red cells to serve as a longer-term source of oxygen during dives when it is not breathing in fresh air.

An inverse correlation is identifiable (and also intuitively anticipated) between the MCV and the erythrocyte count throughout the inframammalian and mammalian phylogenetic taxa, a relationship that was first described in Wintrobe's classical report (1933) and that has been verified by many other workers. That is, the smaller the volume of a given species' individual erythrocytes the greater is the likelihood of the species maintaining a reciprocally higher total number of erythrocytes per unit volume of blood. This relationship is apparent when comparing the erythrocyte counts and the MCVs of the monkey and man.

The mean cellular volume of the rhesus monkey's erythrocyte is acknowledged to vary under multiple normal physiologic and pathologic conditions. The most recognized occurrence of an enlarged red cell (increased MCV) is the polychromatophilic erythrocyte. This is the status of the erythrocyte immediately following the extrusion of the nucleus from the late normoblast. This cell still has some residual ribosomes that, when present in sufficient quantity, impart to it the quality of polychromasia in Wright-stained films of blood and bone marrow. Polychromatic erythrocytes along with nonpolychromatic Rbc that nevertheless still contain some ribsomes are also identified as the precipitate-containing red cells derived by supravital staining with stains such as new methylene blue and brilliant cresyl blue. The polychromatophilic erythrocyte becomes reduced in volume as it evolves into a definitive, mature-sized erythrocyte. It loses the residual ribonucleoprotein, any concomitant mitochondria, and perhaps some other persisting remnant organelles. As would be predicted, it loses its characteristic staining and then has the tinctorial appearance of a typical red cell. Polychromatic erythrocytes are recognized in Wright-stained smears by their off-red, slightly blue-grayish staining, and their often subtly larger size in comparison with accompanying smaller mature red cells. (The term polychromasia refers to the fact that the cell stains with both the methylene blue and eosin of the Romanowsky-type stain mixture thereby giving it a not quite pure red color.) The enhanced size of these cells is the basis for their description as macrocytic red cells (macrocytes). A common pathologic presentation of erythrocytes in the circulation with an increased MCV and polychromatic staining is in hemolytic anemias manifesting a robust erythropoietic response of the bone marrow. Megaloblastic anemia (discussed in detail elsewhere) results in a macrocytosis (classically macro-ovalocytosis) that may or may not demonstrate polychromasia.

Newborn monkeys can be expected to have erythrocytes in their circulation that have a greater MCV than that of the red cells of a normal adult. This is due to the fact that some of the erythrocytes are persistent members of the embryonic hepatic generation of red cells that are likely to be larger than mature red cells derived from the bone marrow in postnatal life as well as the presence of a high number of polychromatophilic (immature) erythrocytes. Umbilical cord blood obtained from newborns at cesarean section performed 10 days prior to predicted labor exhibited Rbc with a definitely enlarged MCV of 94–96 fL (n = 10♂, 7♀, Rogers et al. 2005). The MCVs were electronically determined.

As in man, iron-deficiency anemia in the monkey causes a microcytic anemia (i.e., erythrocytes with a reduced mean corpuscular volume). According to Bicknese et al. (1993, Table 1), iron deficiency in a weanling *Macaca mulatta* causes a statistically significant microcytosis. The MCV of his iron-deficient weanlings was 65 fL (n = 27) while the normal infants' erythrocytes had an MCV of 70 fL (n = 116).

As discussed previously, some species normally maintain different-sized red cells in the males and females. Erythrograms obtained from adult rhesus monkeys in 11 separate investigations were evaluated in order to determine whether a sex-related relationship was identifiable in the MCV of this monkey (Table 1, Switzer 1967a[4]; Robinson and Ziegel 1968[3]; Stanley et al. 1968[5a,c]; Lewis 1977[48]; Matsumoto et al. 1980[8]; Rosenblum et al. 1981[16]; Kessler et al. 1983[25]; Buchl et al. 1997[32]; Andrade et al. 2004[1]; Hassimoto et al. 2004[42]; Chen et al. 2009[33]). The analyses were selected to be comparable and thereby allow the derivation of overall mean values. Representatives of both sexes were analyzed in each experiment. The analyses employed contemporary electronic techniques for the determination of the erythrocyte counts and hemoglobin concentrations of the blood. The hematocrits were determined by microhematocrit centrifugation in the first five of the listed studies and by electronic calculation in the remaining six inquiries. The weighted mean corpuscular volume derived by electronic methods was essentially equal for the sexes. The MCV for males was 72.4 fL (n = 138) while that for females was 73.6 fL (n = 124). On the other hand, the weighted MCVs obtained by employing manually derived microhematocrits in the calculation of the mean cellular volume (as per the Wintrobe formula) were larger, ♂79.3 fL (n = 121) and ♀82.3 fL (n = 154). Thus, the mean cellular volumes based on centrifuged microhematocrit values, at least as derived in this modest meta-analysis, were enhanced by approximately 10% of the levels obtained with electronic calculation. The difference is not unanticipated since it is recognized that a minimal amount of plasma remains in between adjacent erythrocytes in the column of centrifuged red cells in manual determination of the hematocrit resulting in a slightly higher hematocrit than if the trapped plasma were absent. This error is not obtained in the instrumental determination of the MCV because the volume of individual red cells is directly determined by the instrument that subsequently computes a mean cellular volume.

The investigation of Matsumoto et al. (1980) that analyzed six male and female *Macaca mulatta* was the only study of the previous survey that derived statistically significant different MCV values for its males and females (♂77 fL, ♀74 fL). They were obtained by electronic (Coulter counter) determination. Ketamine sedation was not employed.

It is generally recognized, as noted above, that monkeys' erythrocytes are biconcave discs (as are human and most other mammals' red cells). It is noteworthy from a phylogenetic viewpoint that the erythrocytes of all vertebrates beginning with fishes and progressing on through the avians are oval shaped (and also have a permanently residing nucleus). Thus, circular biconcave erythrocytes make their debut in mammals. Gulliver's historic (1875) drawing of the erythrocytes of the various strata of vertebrates illustrates this occurrence (Figure 29). In fact, the very first individual erythrocytes ever to be recognized were those described by Leeuwenhoek in 1668 in his letter to the Royal Society, London. His drawings of a frog's oval erythrocytes whose morphology were accurately perceived by their discoverer are seen in Figure 30. An incidental, intriguing exception to this paradigm are the erythrocytes

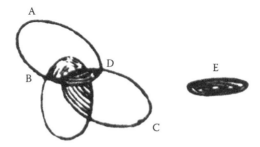

Figure 30   Anton Van Leeuwenhoek's drawing of frog erythrocytes. He accurately perceived their shape and discoidal configuration, describing the red cell "*est particula ovalis sanguinis*" (i.e., it is an oval particle in the blood). (From his letter of July 26, 1684 and also his text Opera Omnia, 1722, *Seu Arcana Naturae*, Leiden, the Netherlands, p. 54.)

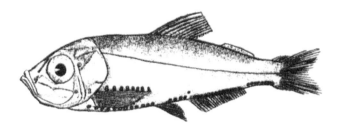

Figure 31 *Maurolicus mulleri.* A fish that has solely denucleated erythrocytes in its blood. It is a small oceanic fish that resembles a miniaturized herring. It is characterized by large, silvery eyes and luminous surface organs. The latter structures would suggest that it is a deep-water fish.

of the deep-sea bristlemouth or light fish *Maurolicus mulleri* (Figure 31). It is a small pelagic (inhabiting open oceans) fish that resembles a miniaturized herring. It is seldom seen by nonichthyologists because of its deep habitat, small size, and fragility. This fish is likely to be observed in plankton-net collections and in the stomach contents of deep-sea predators such as albacore and cod. It is 5–6 cm long and characterized by large silvery eyes and luminous surface organs (Figure 31). Its erythrocytes are small (7 × 2.5–3 µm), ellipsoidal, decidedly biconvex, and (remarkably) denucleated (Figure 32, Hansen and Wingstrand 1960). The MCV is approximately 35 fL, indeed diminutive, that is, about one half of the average volume of the rhesus monkeys erythrocyte. Such cells might be considered as "ahead of their time." The basis for the appearance of denucleated erythrocytes at this early evolutionary level is not understood.

The regular occurrence of oval-shaped erythrocytes in inframammalians (fishes, amphibians, etc.) as opposed to the circular biconcave erythrocytic discs of mammals (including the monkey and man) merits some consideration. These two decidedly different configurations of erythrocytes may be more closely related than initially recognized. It is routinely accepted that the oval configuration for the red cell was evolutionarily selected because it fostered its efficient, frictionless, and turbulence-free flow in blood. In fact, some of the phylogenetically very earliest hemoglobin-bearing cells found in the occasional, scattered invertebrates that have cells containing this respiratory pigment maintain them as free, fluid-suspended spherical coelomocytes, that is, erythrocytes that are spherical (e.g., the polychaete annelid *Pista pacifica*, a bristly multisegmented marine worm). On the other hand, the biconcave erythrocyte of man, monkeys, and other mammals is cited as presenting mathematically documented geometry that maximizes rapid, efficient gaseous exchange. The substitution of the biconcave disc for the oval red cell has been considered a positive biological selection for species with a high metabolic rate and a large oxygen requirement. However, the normally biconcave, discoid erythrocytes of man are seen to assume a reversible ellipsoidal configuration when submitted to shear stress, a condition that mimics flow in a large blood vessel. It is consequently seen that man and presumably the monkey and other mammals have not discontinued the use of oval red cells in favor of the biconcave configuration but rather utilize both forms for the red cell.

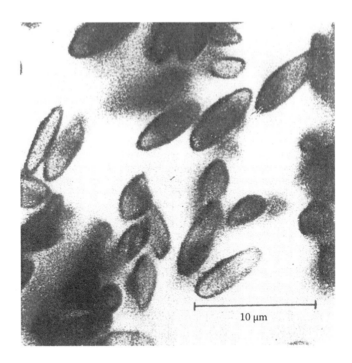

Figure 32 Erythrocytes of the oceanic fish. *Maurolicus mulleri*. The cells are seen to be
denucleated, ellipsoidal, and biconvex. Section of a blood vessel. Bouin fixation,
hematoxylin, and eosin stain. (From Hansen, V.K. and Wingstrand, K.G., Dana
Report No. 54, Carlsberg Foundation, Copenhagen, Denmark, 1960.)

The cell adopts the efficient rheological ellipsoid format when it is being transported
to various organs during laminar flow and when the flow is slower as in a capillary the
erythrocyte "bounces back" to the biconcave configuration that is beneficial for the
exchange of gases. The realization of this functional bimodal reversibility has helped
in the understanding of the genetic condition hereditary elliptocytosis in man (a condition not recognized in monkeys). In this circumstance, the erythrocytes exhibit a
flawed constructed spectrin in the erythrocyte's cytoskeleton. The development of the
normoblasts in the bone marrow is visually normal. The normoblasts have a round
nucleus in a typical-appearing, round cell. However, the red cells in the blood are seen
to be oval in shape. It is thus reasoned that such red cells normally discharge their
nucleus in the marrow and enter the blood as customary-configured erythrocytes.
When a given red cell enters the laminar, rapid flow of a muscular artery, for example,
the red cell assumes the oval shape. Then following an unspecified number of cycles
between oval and discoid shapes, the red cell is unable to return to the biconcave disc
shape because of its defective cytoskeleton. It thereby thereafter permanently retains
the oval shape. Such relatively rigid red cells are less capable of negotiating through
tight restricted spaces such as those in the spleen. They sustain increased trauma
under various conditions and have an abbreviated life span. Depending on the severity

of the disease, this can lead to the requirement for normoblastic hyperplasia and in some cases a hemolytic anemia develops.

The second erythrocyte index is the mean cellular hemoglobin (MCH) and indicates the average quantity of hemoglobin in an individual erythrocyte. It is expressed as picograms (pg) of hemoglobin per cell. A strong positive correlation exists between the size of the erythrocyte (MCV) with the amount of hemoglobin it contains; for example, in birds r = 0.90, P < 0.001 (Hawkey 1991). Since the magnitude of a red cell can vary widely, so can its total content of hemoglobin. The range for submammalians extends from ~30 to >3000 pg of hemoglobin per cell. The grand weighted MCH for the *Macaca mulatta* that were analyzed solely by electronic methods cited in the prior discussion of the MCV (six separate experiments, n = 138♂ and 124♀) is 22.6 pg for the males and 22.8 pg for the females. The electronically determined MCH for the larger human erythrocytes is in the range of 26–34 pg (Kratz et al. 2004). It has been noted that in man when an anemia is developing, the fall in hemoglobin concentration of the blood can sometimes be proportionally greater than the decrease in the number of erythrocytes and consequently yield a bona fide slightly lower MCH. This could be interpreted as a combination of a quantitative numerically reduced production of red cells and a decreased proficiency in their construction. Presumably, this phenomenon may also be identifiable in some anemic conditions in monkeys.

The third erythroid index is the mean cellular hemoglobin concentration (MCHC) and it can be considered as denoting the percentage of an average erythrocyte that consists of Hb. This indicator is derived from the formula: grams of Hb/dL of blood × 100 divided by the hematocrit expressed as a whole number, for example, 45. Thus an MCHC of 32 indicates that the concentration of hemoglobin in an average cell is 32%. A trend for progressively higher intraerythrocytic concentrations of hemoglobin is seen to accompany the evolutionary advance of the submammalians including the birds (Glomski and Pica 2011). Wintrobe (1933) demonstrated that of all the red cell–dependent values derived from the hemogram (Rbc, Hct, Hb, and erythrocyte indices), the MCHC maintains the greatest constancy throughout the phylogeny of the vertebrates (Figure 33). The limited species to species variation of the MCHC is appreciated partly in the recognition that since hemoglobin is a large complex protein molecule, it is likely to have a limited range of solubility. Since hemoglobin that is contained in the cytoplasm of an erythrocyte is present in a relatively high concentration, typically 25%–35%, the solubility characteristics of this molecule self-limit the potential of significantly increasing its intraerythrocellular concentration. (The hematocrit is the next relatively least variable erythrogramic parameter of the vertebrate subphylum [Figure 33].)

The mean cellular hemoglobin concentration (MCHC) that is calculated with the value of a centrifuged hematocrit in the Wintrobe formula is a less precise value than obtained with an electronically configured hematocrit.

The electronically determined weighted MCHC for the rhesus monkey (derived from the same group of six individual studies that were cited for the computation of the weighted MCH (n = 138♂ and 124♀) is 31.6% for males and 31.2% for females. Thus, it is seen that although the red cell count of the monkey is greater than that of man, and the mean cellular volume of the *Macaca mulatta* erythrocyte

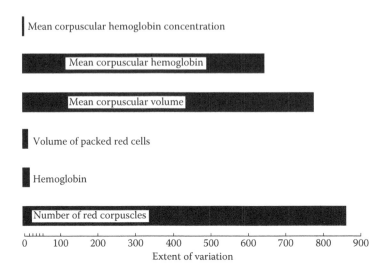

Figure 33   Wintrobe's illustration of the relative magnitude of the differences in erythrocyte-based values across the spectrum of vertebrates (both mammalians and infra-mammalians). It can be seen that the erythrocyte count has the greatest range of difference among the different species of vertebrates, while the mean concentration of hemoglobin in the red cell (MCHC) demonstrates the least variance from species to species, that is, this quantifiable characteristic is the most constant and uniform among all animals' red cells. (From Wintrobe, M.M., *Folia Haematol.*, 51, 32, 1933.)

is considerably smaller than man's red cell, the MCHCs of man and the monkey are of the same order (man, 31%–37%, Kratz et al. 2004). This close agreement under-scores Wintrobe's finding (1933) that the MCHC is the least variable of erythrocytic parameters throughout the vertebrates (Figure 33). One observation that reinforces the concept that the MCHC is a particularly stable value is obtained from eryth-rograms obtained from *Macaca mulatta* umbilical cord blood at cesarean section (Rogers et al. 2005). Ten males' and seven females' hemograms were analyzed (Rbc, Hct, Hb, and erythrocytic indices). It was seen that even though the mean red cell counts were markedly lower than what is seen in the adult ($\male$ infant 4.18 × 10⁶/μL, $\female$ infant 4.35 × 10⁶/μL) and the MCV was comparably elevated ($\male$94 fL, $\female$96 fL), the MCHC was nevertheless the standard value ($\male$32.8%, $\female$32.5%).

The one specific condition in man in which the MCHC is verifiably (electroni-cally) increased is hereditary spherocytosis (>36%). This disease, it appears, has not been identified in monkeys.

The weighted erythrocyte MCV of 3–5-year-old male and female *Macaca fascicularis* derived from phlebotomies that were conducted while the animals were alert and not administered any sedation is equal for both sexes, a weighted mean 80–81 fL (n = 105$\male$, and 215$\female$, Table 5). Five- to 6-year-old subjects gener-ate similar weighted mean values, males 77 fL and females 78 fL (n = 52$\male$, and 55$\female$, Table 5). In comparison the weighted mean value for ketamine-administered subjects 3–5 years old was 78 and 75 fL for the males and females, respectively (n = 44♂, and 41♀, Table 6). The mean cellular volumes in nonsedated monkeys were quite consistently similar in both sexes regardless of the age of the subjects (Table 5). On the other hand, the values were more varied in paired groups of sexes that were anesthetized with ketamine prior to venisection. There is no indication, however, that the presence or absence of the medication is associated with significant different red cell volumes. There are also no definite gender-based differences in mean red cell volume. The range of MCV of nonmedicated monkeys is seen to be rather narrow, that is, 78–81 fL (Table 5). The range in ketaminized cynomolgus monkeys, however, is more heterogeneous. The level of 66 fL obtained in 1- and 2-year-old treated monkeys seems rather different from the 78 fL observed in unmedicated monkeys (Tables 5 and 6). The 68 fL level obtained in 5–10-year-old females and 8–18-year-old males is not readily explained. The MCVs for the rhesus and cyno-molgus monkeys obtained with contemporary electronic investigation are thus seen to be of the same order, approximately 70–80 fL, and definitely smaller than man's red cells.

As noted previously, Xie et al.'s (2013) and Yoshida et al.'s (1986a) independent years—separated examinations of large groups of *Macaca fascicularis*—revealed a tendency for the MCV to increase in both sexes during the 2–5 years of age interval (Table 1, Figure 27). It attained statistical relevance in Yoshida's monkeys.

The weighted mean cellular hemoglobin (MCH) values for the age–sex grouped *Macaca fascicularis* whose erythrograms were derived from the blood samples obtained without the use of ketamine were consistently 22–23 pg (Table 5). The comparable values for the cohorts that were sampled from ketaminized monkeys were in a similar range (Table 6). However, animals that had generated smaller MCVs in the 66–68 fL span tended to yield lower MCH levels (Table 6). All members of each ketaminized and non-ketaminized age–sex subsets of monkeys listed in Tables 5 and 6 presented a weighted mean cellular hemoglobin concentration of 28%–29%. The one exception was the 8–18-year-old males administered ketamine who yielded a weighted MCHC of 32%.

CHAPTER 14

Erythrogram
Variability Associated with Age

Adolescent female rhesus monkeys (like human females) show a growth spurt followed by menarche (initiation of menses), then a period of continuing growth, intense bone mineralization culminating in full sexual maturity, and cessation of linear growth (Golub et al. 1999). This stage prevails during 24–48 months of age in female *Macaca mulatta*. Changes in the blood picture could logically be conjectured to occur at this time. Golub et al. (1999) conducted an analysis of the erythrocellular profile of healthy prepubertal female rhesus monkeys during this period, that is, after their adolescent growth spurt had begun and about time of onset of menses. The subjects were members of the colony maintained at the California Regional Primate Research Center, University of California, Davis, and their mean age was 29.6 months. Some animals were placed on a control diet (n = 8) while others were assigned to a diet that was mildly deprived of iron and zinc (n = 16). The normal control diet delivered an estimated 2.9 mg Zn and 1.7 mg Fe per day. The deficient diet was assessed to supply 0.2 mg Zn along with 0.8 mg Fe per day. Blood samples for hematologic analyses were obtained at the time of initiation of the dietary interventions and 3 months later at the end of the growth spurt. During this 3-month growth spurt interval, the normal control monkeys gained 7% of their baseline body weight/month. Linear growth was also evident. The erythrocyte count, hematocrit, and hemoglobin of both control and iron- and zinc-deficient groups significantly declined during the 3-month test period (Table 1). Of potential consideration was the fact that both cohorts demonstrated a small but statistically significant increase in the MCV (reflecting an increase in reticulocytes?). The primary erythroid parameters persisted in the range obtained at the end of the third month through to the end of an additional 3 months of study for both groups (termination of the experiment). It was seen that the hematologic picture of the cohort maintained on the mildly deficient diet was not significantly affected but the plasma iron and zinc were somewhat lower. However, four of the eight deprived monkeys had an iron-deficiency anemia as opposed to none of the controls at the end of the total 6 months of experimentation.

The Primate Aging Database, a multicentered database, was established for nonhuman primates including *Macaca mulatta* (Smucny et al. 2001, 2004). The animals were maintained at three institutions: (1) the National Institute of Aging

Research Group at the National Institutes of Health Animal Center, Poolesville, Maryland, (2) the Wisconsin Regional Primate Research Center, Madison, Wisconsin, and (3) the Oregon Regional Primate Research Center, Beaverton, Oregon. The aggregated data were applied to the examination of the age-related hematologic (and biochemical) changes across the adult life span of the rhesus monkey. The information was obtained from healthy, primarily indoor-housed animals that were regularly sampled at health assessments over several years. All participants had known dates of birth. Data were analyzed from subjects 7 to 36 years old. Seven years of age was selected for the lower limit of adulthood for the rhesus monkey because it represents the median age between the onset of puberty (usually by the age of 4 years) and the completion of developmental processes such as bone maturity (by the age of 10 years). Blood samples were drawn either without sedation or following ketamine injection (7–10 mg, IM) in the morning after an overnight fast. The erythrograms were based on a population of ~240 monkeys with ~3000 samplings for each erythroid parameter (Rbc, Hct, Hb, and erythrocyte indices).

The data derived from these studies were analyzed in multiple ways (Smucny et al. 2001, 2004). The simplest format was the determination of the mean primary erythroid values for the monkeys at the age of 7 years (Rbc 5.29 × 10⁶/μL, Hct 40.4%, Hb 12.8 g/dL) and the mean for the interval of 6–36 years (Rbc 5.59 × 10⁶/μL, Hct 41.4%, Hb 13.3 g/dL) (Table 1 and Table 11). Higher values than those obtained for the "basal" 7-year-olds are seen to have developed during the extended monitoring period. Scatterplots were prepared for each erythroid parameter and included a calculated regression line. In addition, plots of predicted means for each year of age were also developed using the coefficients generated from multiple linear (quadratic) regression models for age. Thus, Smucny et al. (2001, 2004) identified a sex-neutral pattern

Table 11   Erythrocyte-Related Values for *Macaca mulatta* 6–30 Years of Age
Developed by the Primate Aging Database[x]

	Mean at Age 7 Yr	Mean for Ages 6–36 Yr	P-Value, Effects of Age on Parameter	Ages (Years) with Mean Values 1 SD Different from Age 7, Arrow Indicates Direction of Change*
Erythrocytes	5.29	5.59	0.001	20–31*↑
Hematocrit	40.4	41.4	0.007	23–26, 30↑
Hemoglobin	12.8	13.3	0.01	22–26, 28, 30↑
MCV	76	75	<0.001	27–31↓
MCH	24	24	0.003	29, 31↓
MCHC	—	32	—	—

Sources: The superscript reference numbers refer to the identity of the investigators in Table 1.
Notes: Erythrocytes millions/mm³ or μL, Hct, %; Hemoglobin, g/dL; MCV, mean cellular volume (μm³ or femtoliters (fL)); MCH, mean cellular hemoglobin (pg); MCHC, mean cellular hemoglobin concentration (wt./vol%, g Hb/100 mL Rbc).
* For this parameter and this age group the 30-year-old subjects' data demonstrated 2 SD. n = 231–246 for the 6–30-year-old age range subjects with 2952–3093 samplings per parameter. The data of the table are sex neutral. Smucny et al. 2001[56], 2004[56, x].

of the erythroid blood picture for the adult years of *Macaca mulatta*. As reported, some of the monkeys were treated with ketamine while others were not. Assuming the proportion of medicated subjects remained constant in the investigation, the profiles of the curves should not have been affected (even though the actual erythrocyte count levels, and so on, would be expected to be modified due to the analysis of a mixture of ketaminized and non-ketaminized subjects). While as indicated, the calculations were controlled for sex, sex-specific profiles though planned to be derived have yet to be presented. Analyses of the assembled data revealed that each of the erythrocellular-based values exhibited a statistically significant age-associated change (Table 11). The single exception is the MCHC. This latter stability is seemingly in accord with the finding of Wintrobe (1933) that the MCHC, from an overall vertebrate phylogenetic perspective, is the least variable of the erythrogramic values (Figure 33). The graphs for the derived predicted mean values with age developed by Smucny and her fellow workers are particularly helpful in visualizing the general direction of the levels of the parameters throughout the studied segment of adult life (7–30 years). It is interesting to observe that all of the major erythroid values (i.e., Rbc, Hct, Hb) displayed a gradual, slight (in the case of the Hct and Hb), or modestly greater (in the case of the Rbc count) enhancement with age. Conversely, the MCV and MCH had a minimally corresponding diminution. The magnitude of change between the mean value at 7 years of age versus the mean value for the interval of 6–36 years of age was an increase of 5.7%, 3.9%, and 2.5% for the Rbc, Hb, and Hct levels, respectively, while the decrease for the MCV was 1.3% and that of the MCH was 1.2%. Further dissection of the erythrogramic profiles established the ages in which the mean values differed one standard deviation from the level observed at the age of 7 years (Table 11). These specific identifiable points occurred during the relatively older ages of 20–31 years (Table 11).

Erythrograms of monkeys of specific ages (newborns to older than 10 years of age) from comparable experiments are presented in Table 12. Data from equivalent studies for monkeys of the same age and sex are expressed as weighted mean values. The results of monkeys administered ketamine and those not given the sedative are reported separately. Examination of the age-correlated mean values presented in the table identifies some facets regarding the rhesus blood picture. The blood of newborn infants (obtained from the umbilical cord blood at the time of cesarean section 10 days prior to expected delivery) presented a decidedly low erythrocyte count. These red cells had a very large MCV and were a population presumably maintaining fetal hemoglobin. The following listed cohort was comprised of approximately 3-month-old weanling infants that displayed an Rbc count that was increased over that of the cited newborns but was still not as high as seen in most monkeys of older age. The MCV of their red cells was reduced to "normal" (70 fL) (Bicknese et al. 1993). This reduction in size most likely reflects the entry of (smaller) definitive generation erythrocytes into the circulation and perhaps also a decrease in the proportion of reticulocytes (which are larger than fully mature Rbc) in the blood. In most instances, the values for the Rbc, Hct, and Hb of the different age groups are minimally higher in males. Monkeys of a given age and sex that were administered ketamine typically have lower erythrogramic values than counterparts that were not given the sedative. In terms of the level of the erythrocyte count for sexually mature

Table 12 Erythrograms: Age Groups, Sex, and Presence or Absence of Ketamine in Institutional Captive *Macaca mulatta*

	Rbc	Hct	Hb
Newborn infants at cesarean section, umbilical cord blood			
Rogers et al.[47]			
n = 10♂, MCV 94 fL	4.18	39.9	13.1
n = 7♀ MCV 96 fL	4.35	41.8	13.6
Weanlings, ~3 months			
Bicknese et al.[39]			
n = 116, mixed sex, no anesthetic	5.28	37.0	12.0
Age: 3 weeks to 1 year			
Buchl et al.[32]			
n = 27♂ with ketamine	6.18	41.4	13.1
n = 27♀ with ketamine	6.04	40.4	12.8
Age: 0.8–1.2 years			
Fernie et al.[34]			
n = 5♂, 4♀, no anesthetic	5.71	41.0	13.2
n = 5♂, 4♀, with ketamine	5.43	38.0	12.4
Age: 1–2 years			
Buchl et al.[32]			
n = 30♂ with ketamine	5.73	40.0	12.8
n = 77♀ with ketamine	5.75	40.1	12.9
Age: 2–3 years			
Buchl et al.[32]			
n = 27♂ with ketamine	5.71	39.7	13.0
Lilly et al.[24], Buchl et al.[32]			
n = 95♀ with ketamine	5.41	39.9	12.6
Weighted mean values			
Stanley et al.[5b]			
n = 36♂ no anesthetic	5.70	—	14.1
n = 22♀ no anesthetic	5.57	—	13.4
Age: 3 years			
Woodward et al.[19]			
n = 9♂, 9♀, with ketamine	5.16	36.4	11.5
Age: 3–5 years			
Stanley et al.[5c]			
n = 38♂ no anesthetic	5.85	—	13.8
n = 34♀ no anesthetic	5.60	—	13.5
Age: 3–5 years			
Hom et al.[43]			
n = 53♂, with ketamine	5.32	38.9	12.6

(*Continued*)

Table 12 (*Continued*)   **Erythrograms: Age Groups, Sex, and Presence or Absence of Ketamine in Institutional Captive *Macaca mulatta***

	Rbc	Hct	Hb
Age: 4–10 years			
Switzer et al.[4], Matsumoto et al.[8]			
n = 11♂, no anesthetic	5.91	46.6	13.9
Weighted mean values			
Switzer[4], Matsumoto et al.[8]			
n = 26♀, no anesthetic	5.85	44.5	14.0
Weighted mean values			
Kessler et al.[25], Buchl et al.[32]			
n = 75♂, with ketamine	5.91	42.4	13.4
Weighted mean values			
Kessler et al.[25], Buchl et al.[32]			
n = 67♀, with ketamine	5.79	42.0	13.1
Weighted mean values			
Age: Older than 10 years			
Buchl et al.[32], Bennett et al.[44]			
n = 44♀, with ketamine	5.84	41.8	13.4
Weighted mean values			
Bennett et al.[44]			
n = 15♀, no anesthetic	5.97	44.7	14.1

Sources: The superscript reference numbers refer to the identity of the investigators in Table 1.

Notes: Mean values, Rbc = × 10^6/µL; Hct = %; Hb = g/dL. Erythrograms were analyzed by contemporary electronic techniques. Manual hematocrits were not included. As indicated, data for groups of subjects derived from more than one investigation are reported as weighted mean values.

monkeys (≥4 years) it is seen that the mean approaches 6.0 million/µL but rarely exceeds this level. It is noted, however, that the number of monkeys evaluated in Table 12 older than 10 years is limited. No males are included and definitive elderly (postmenopausal) females are not indentified. The data of Smucny et al. (2001, 2004) (Primate Aging Database) (Table 11) indicate increases in the Rbc count occurring at the age of 20 years and later, extending on to 31 years. The mean erythrocyte count of 5.59×10^6/µL reported by Smucny and her coinvestigators for the more extended period of 6–36 years is analytically consistent with the shorter-term results presented in Table 12.

The occurrence of menopause, the cessation of menstruation, appears to be the most universal phenomenon with the least variation in age in female rhesus monkeys (as well as in women) (Uno 1997). Other though less common age-related biomarkers are coronary artery sclerosis, degenerative joint disease, and cancer. It is generally agreed that by the age of 20 years reproductive activities become reduced and menopause occurs at 26–27 years (Walker 1995). The number of annual menstrual cycles begins to decrease after the age of 22–25 years. Major geriatric pathology starts to develop in the rhesus macaque by 20 years of age, approximately 5 years before menopause. The average life expectancy of captive female *Macaca mulatta* appears to be several years after menopause. Whether there are any specific erythro-cellular correlations with menopause has not been given unambiguous investigation. Smucny et al.'s (2001, 2004; Table 11) sex-neutral erythrogramic data, discussed previously, indicate that there are differences of one standard deviation between the mean levels of the major erythroid values (Rbc, Hct, Hb) obtained at 7 years of age versus the levels observed during the premenopausal and menopausal years suggesting at least a coincidence of these hematologic and reproduction-related changes.

CHAPTER 15

Effects of Fasting on the Erythrogram of *Macaca fascicularis*

The effects of fasting on the erythrogram of 4–5-year-old cynomolgus monkeys (n = 6♂, 3.1–6.0 kg; n = 5♀, 2.8–3.6 kg) were assessed by Zeng et al. (2011, Table 1). Blood samples were obtained from the cephalic vein of conscious subjects who were fasted for 0, 8, 16, and 24 hours. All phlebotomies were obtained at the same point in time on different days. At least 24 hours elapsed between each of the phlebotomies. The monkeys served as their own controls. Significant differences were observed between the erythrograms of male monkeys obtained at 0 hours of fasting and 16 hours of fasting. The Rbc counts, Hct, Hb, and MCV levels were all significantly lower and the MCH and MCHC were significantly higher in the 16-hour fasting samplings (Table 1). Although significant differences were not attained following 8 hours of fasting (except for the MCH), all major erythrocytic parameters (Rbc, Hct, Hb) displayed a mathematical direction toward the values derived at 16 hours of fasting. Thus although these "modifications" did not attain significance, a bias toward the 16-hour level was present. On the other hand, no substantive changes were obtained at 24 hours when compared with the 0-hour (nonfasting) levels. However, the values at 24 hours of fasting for each of the three primary erythroid factors showed a numerical tendency to recover some of the decrement that had occurred at 16 hours of fasting. That is, the values though still less than that obtained at 0 hours of fasting were minimally higher than those at 16 hours of fasting.

In regard to the females in this study, they consistently presented lower values for all erythroid parameters (Rbc, Hct, Hb, and erythroid indices) than the males for all studied intervals (nonfasting and 8, 16, and 24 hr of fasting) (Table 1). The only minor exception was the equal MCHC (33.5%) obtained at 16 hours of fasting. Further, in contrast to the males, the females did not present any significant differences in the level of erythroid parameters between the nonfasting state and any of the three periods of fasting. However, there was a definite tendency for the females' erythroid values to maintain the same fasting-associated direction of change maintained by the males. But it never reached the level of significance. Thus at 16 hours of fasting, the interval of greatest change for the males who displayed significant decrements in Rbc, Hct, Hb, and MCV in comparison with 0 hours of fasting, the females offered a change in the same direction but of lesser proportions. Also just as the males minimally recovered at 24 hours of fasting some of the losses observed at 16 hours of fasting in the levels of Rbc, Hct, and Hb, the females also showed equivalent nonsignificant higher values than those of the 16-hour sampling.

This investigation is limited by its modest number of subjects but nevertheless has a sufficiently large "n" to permit statistical analysis of the findings. It fosters comparisons between the sexes and yields information that helps support some of the basic concepts concerning the erythrogram of the long-tailed macaque and possibly the rhesus monkey. The results of the investigation are in accord with the tenet that fasting and nonfasting erythrograms are not equivalent. This study suggests that the most significant differences in fasting and nonfasting erythrograms of the cynomolgus monkey are obtained at 16 hours of fasting in males. The mean major alteration in erythroid parameters (Rbc count, hematocrit, and hemoglobin concentration) seems to be a decrease of the values in these parameters at 16 hours of fasting (in comparison with nonfasting levels). It is interesting to observe that the mean fasting-associated alterations of the erythrogram of female cynomolgus monkeys tend to parallel the modifications presented by males of this species. Thus, the direction of alterations obtained at 16 and 24 hours of fasting is qualitatively similar to those of males but of lesser magnitude. This small survey also supports the concept that the cynomolgus males normally maintain higher erythrocyte counts, hematocrits, and hemoglobin concentrations than that of females.

CHAPTER 16

Erythrogram of Infant *Macaca mulatta*

While it is recognized that various circumstances such as lifestyle, diet have an impact upon the erythrocyte count and other erythrocyte-dependent values of mature monkeys, certain factors also moderate the hemogram of the infant *Macaca mulatta*. In one examination of this aspect of hemopoiesis, blood samples were obtained from mother-reared (n = 6) and nursery/peer-reared (n = 17) laboratory-housed rhesus macaque infants on days 14 and 30 and during months 2, 3, 4, and 5 of their life (Kriete et al. 1992). The mother-reared infants consistently exhibited higher erythrocyte (and lymphocyte) counts across all time points. Hemoglobin and hematocrit values were similar in both groups on days 14 and 30 and month 2. However, these values were higher in the nursery/peer-reared infants at months 3, 4, and 5. The Rbc distribution width of the nursery/peer-reared group decreased and was lower than that observed in mother-reared infants during months 3, 4, and 5 (an indication of improved health of the former group?). This latter outcome concerning the nursery/peer set along with a concomitant increase in this cohort's MCV would suggest, among other choices, that the production of their erythrocytes was becoming more efficient than it was during earlier life. The presumed advantage of a higher erythrocyte count obtained by mother-reared infants is an interesting phenomenon that stimulates further consideration.

The hematocrits (per microhematocrit technique) of mixed-sex neonatal, laboratory-bred, hand-reared *Macaca mulatta* were monitored for 2 years postdelivery by Martin et al. (1973, Figure 34). Samples were obtained once during the first 3 days of life, at 1, 2, 3, 4, and 8 weeks, and monthly during the remainder of the first year. During the second year, the subjects were phlebotomized at 6-month intervals. One hundred sixty-eight infants were analyzed perinatally and more than 100 were studied through 18 months. Thirty-seven monkeys were evaluated at 2 years of age. One would assume that it would not be necessary to divide the sexes into two groups as a sex-linked relationship with the hematocrit would unlikely be extant at this early age. This longitudinal study offers a graphic profile of the level of the hematocrit from birth onward for 2 years. It reveals that the hematocrit was markedly elevated during the first 3 days of life (newborn Hct ~ 50% versus adult Hct ~ 44%). The level immediately precipitously declined so that at 2 weeks it was lower (Hct ~ 37%) than the mean of the adult (Figure 34). This was followed by an increase that attained 39%

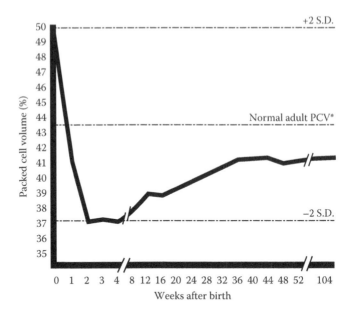

Figure 34 *Macaca mulatta*. A graph illustrating the changes of the hematocrit of a neonatal, laboratory-bred, hand-reared *Macaca mulatta*, mixed-sex population, from birth to 2 years of age. The values were quantitated by the microhematocrit technique. Hematocrits were determined once during the first 3 days of life, at 1, 2, 3, 4, and 8 weeks, and monthly the rest of the first year. During the second year, values were assessed at 6-month intervals. There were 168 infants that were analyzed at birth, and at least 100 were studied through 18 months. There were 37 monkeys that were phlebotomized at 2 years of age. The hematocrit was definitely elevated at birth but declined rapidly so that by the second week it was lower than the mean mixed-sex value for the adult. This was followed by an increase, which attained 39% at 12 weeks, and 42.7% at 2 years. The dotted line at the mid-height of the graph (hematocrit ~43.5%) indicates the normal adult mean (for the mixed-sex population), while the highest and lowest dotted lines indicate the range for ± 2 SD from the mean adult value. The y axis indicates the range of hematocrit values from 35 to 50, while the x axis indicates the duration of time in weeks from zero (day of birth) to 104 weeks. (From Martin, D.P. et al., *Lab. Anim. Sci.*, 23, 194, 1973.)

at 3 months and ~42% at 2 years. Conversely, it is seen that the newborns assayed by Rogers et al. (2005, Tables 1 and 12) do not present a markedly high hematocrit as observed in the (previous) newborns studied by Martin et al. (1973, Table 1). This disparity is reconciled by the fact that the two populations have specific differences. Rogers' infants were analyzed immediately at cesarean delivery, and thus, the infants had an atraumatic delivery and their hemic picture could potentially reflect some effects of the anesthesia administered to the mothers. On the other hand, Martin's subjects not only sustained the repercussions of a regular unanesthetized vaginal delivery but also their initial hematocrit was obtained within 3 days of birth that allowed the physiologic adjustments to birth and onset of extrauterine life.

There is an implication that the erythrogram of the infant monkey aged 3–5 months (12–24 weeks) may present lower erythrocyte counts, hematocrits, and hemoglobin concentrations of the blood than older infants perhaps up to 1 year of age. During the 3–6 months of age interval, Fernie et al. (1994) reported that their unsedated rhesus monkeys displayed median lower Rbc, Hct, and Hb levels than when they were 46, 50, or 58 weeks old and were also analyzed without the administration of ketamine (n = 9, Table 1). The infants were weaned at 22 weeks. In another comparison, normal approximately 3-month-old weanlings (n = 116) that were analyzed in the absence of ketamine presented the relatively low erythrocyte count of 5.28 million/µL (along with Hct 37% and Hb 12.0 g/dL) (Tables 1 and 12; Bicknese et al. 1993). And a group of rhesus monkeys that were sampled over the more extended age period of 3 weeks to 1 year under ketamine sedation, (which would theoretically tend to reduce the erythroid values) yielded higher major erythrocyte values than that of the foregoing 3-month-old weanlings (Tables 1 and 12; n = 27♂, 27♀; Buchl et al. 1997).

No significant different differences were identifiable in the erythrograms of infants delivered from primagravid versus secundigravid dams who can also be classified as younger 4–5-year-old or older 5–6-year-old *Macaca mulatta* (Rogers et al. 2005).

In an assessment of the blood picture of infant rhesus monkey weanlings, it was found that subjects maintained indoors (at the California Primate Research Center) raised entirely or partially in the nursery had a lesser prevalence of iron deficiency (demonstrated by their hemograms) than dam-reared offspring (Bicknese et al. 1993). The results further suggested, not surprisingly, that multiparous dams were more likely to have iron-deficient weanlings than primaparous dams. Iron-deficient weanlings had a statistically significant lower MCV than the healthy controls (65 fL versus 70 fL). In addition to an expected significantly lower hematocrit and hemoglobin level, the deficient monkeys had a significantly higher mean erythrocyte count (5.48×10^6/µL) than the normal subjects (5.28×10^6/µL). It can be proposed that this difference represents a compensatory erythropoietic response on the part of the affected infants. This erythrocytosis is thus one example of significantly different erythrocyte counts among weanlings. The prevalence of iron deficiency in this study was 19%, a rather high proportion it seems, for monkeys reared under modern, ideal laboratory conditions.

CHAPTER 17

Erythroid Profile of Fetal
Macaca fascicularis

The blood picture of *Macaca fascicularis* during its *in utero* fetal development was established by a study of 31 fetuses beginning on fetal gestation day (GD) 80 through gestation day 150 (Table 1, Tarantal 1993). A total of 68 samplings were assessed (2–5 samples per fetus). The blood samples were obtained by ultrasound-guided cardiocentesis from ketamine-immobilized (10 mg/kg, IM) females whose time of conception was systematically attested with a maximum error of 2 days. The length of pregnancy is ~165 days although it has been reported to extend to 190 days. The general pattern of the developing erythroid blood picture initiating on gestation day 80 was a steady increase in the erythrocyte count, hematocrit, and hemoglobin concentration that continued throughout gestation until a slight decline occurred in all three values close to term at about 150 days of pregnancy (Table 1). The earliest erythrogram was Rbc $3.02 \times 10^6/\mu L$, Hct 33.0%, and Hb 11.3 g/dL and by GD 150 the values had attained $4.03 \times 10^6/\mu L$, 36.7%, and 12.4 g/dL, respectively. The MCV demonstrated a gradual decline throughout the period. The earliest value was 109 fL but at close to term (150 days) it was still larger than that of the adult, that is, mean 91 fL versus the usual 70–80 fL obtained in adults. The enhanced size is a reflection of multiple factors including the fact that these are a population of cells derived most likely from the liver as well as the primitive generation that originated in the yolk sac and this pool of cells has a large component of reticulocytes. The MCH demonstrated a parallel declining course illustrating the diminishing quantity of hemoglobin in smaller red cells, while the MCHC offered the consistent stable level that is typical for this parameter. The numerical occurrence of nucleated erythrocytes in the circulation was most interesting and variable. These cells were most numerous at the earliest assessment (GD 80), the mean was ~1000 per 100 circulating leukocytes. Thus, these nucleated cells were more numerous than white cells. The range observed among the subjects (n = 7) was particularly broad, 63–3500/100 wbc. Ten days later the mean dropped markedly to 300 nucleated red cells/100 leukocytes (range 0–1600, n = 12). A further reduction to a mean 63/100 wbc (range 0–300) was obtained at 110 days gestation. They were relatively rare by 140–150 days with an average incidence of 7 normoblasts per 100 leukocytes. Thus, this population persisted as a significant component of the circulating pool of cells throughout pregnancy because even at gestation days 140 and 150, their mean absolute level was 469 and 490 nucleated red cells per microliter of blood. As would be expected, poly-chromasia and anisocytosis were seen throughout gestation. Hypochromasia, poi-kilocytosis, and Jolly bodies were uncommon. Tarantal (1993) has underscored the near unanimity of the fetal development of hemopoiesis in man and the macaques.

CHAPTER 18

Erythrogram of Infant *Macaca fascicularis*

The blood picture of cynomolgus monkeys was monitored on the day of birth, thereafter daily for the first week of life, and then monthly through the 11th month of infancy by Sugimoto et al. (1986, Table 1). The subjects were born and reared at the Tsukuba Primate Center for Medical Science, Japan. They were full term and had a natural delivery. A total of 131 monkeys were used. Five infants were analyzed for each daily study from birth through the first week of neonatal life. The monthly studies included a minimum of 3 and a maximum of 13 monkeys (mean = 8).

The mean primary erythroid values at birth were Rbc $6.65 \times 10^6/\mu L$, Hct 56.8%, and Hb 16.8 g/dL. The MCV was 85 fL (Tables 1 and 5). They were elevated levels. The mean values for the 7 days following birth (n = 5 infants per day) were Rbc $6.55 \times 10^6/\mu L$, Hct 54.1%, and Hb 15.9 g/dL (Table 5). The MCV was 83 fL. A minimal reduction was thus seen in each parameter. By 1 month, a further lower plateau was established: (n = 13), Rbc $5.53 \times 10^6/\mu L$, Hct 41.2%, Hb 12.3 g/dL, and MCV 75 fL (Tables 1 and 5).

The mean values for the successive 5 months (2–6) and the subsequent 5-month period (7–11) are very similar. An aggregate of 78 subjects were analyzed during this period. The average values derived from 37 analyses performed during the second through the sixth month of life were Rbc $6.87 \times 10^6/\mu L$, Hct 45.5%, Hb 11.7 g/dL, and MCV 66 fL. The mean levels for the subsequent 7–11 months yielded the following: Rbc $6.68 \times 10^6/\mu L$, Hct 45.9%, Hb 12.0 g/dL, and MCV 69 fL (n = 41, Tables 1 and 5). Among the observations that can be made in this investigation is that the MCV and MCH progressively decrease from birth through the first year.

Comparison of the hemograms of umbilical cord blood from cesarean-sectioned infants at delivery (n = 6) and femoral blood samples from the same individuals 5 hours after birth revealed that increased levels of the erythrocyte count, hematocrit, and hemoglobin concentration were obtained in the 5-hour samples suggesting a reduction of the plasma volume of blood by an exudation of fluid out of the vessels into the interstitial tissue. The mean values for the first and second samplings, respectively, were Rbc 4.98 versus $5.91 \times 10^6/\mu L$, Hct 44.1 versus 49.9%, and Hb 12.7 versus 14.9 g/dL (Tables 1 and 5). The umbilical cord blood values at cesarean section of the cynomolgus monkeys are consistent with the pattern of similarly delivered

Macaca mulatta (Tables 1 and 12, Rogers et al. 2005). The red cell count is lower in the cesarean-delivered infants than in those who underwent natural delivery. The MCV is higher in cesarean-sectioned macaques of both species presumably due to a higher proportion of fetal hemoglobin-bearing, liver-originated red cells and perhaps a greater incidence of reticulocytes.

CHAPTER 19

Simian Immunodeficiency Virus (SIV)

Its Relationship with Erythropoiesis in Macaca mulatta

The striking similarities between simian immunodeficiency virus (SIV)-induced disease in rhesus monkeys and human immunodeficiency virus (HIV) infection in man make SIV-induced disease in rhesus monkeys a very important model for the study of AIDS (acquired immune deficiency syndrome) in man. The present discussion is designed to specifically address the impact that the simian immunodeficiency virus infection has upon the erythropoiesis of *Macaca mulatta.*

The human and simian immunodeficiency viruses are retroviruses, that is, single-stranded, diploid, enveloped RNA viruses. They have the ability to transfer genetic information in a retrograde fashion from RNA to DNA by virtue of their possession of an enzyme that is an RNA-dependent DNA polymerase termed reverse transcriptase. The virus attaches to the cell membrane of the target cell typically at a specialized receptor site. In the case of simian as well as human immunodeficiency viruses, the receptor is the CD4 receptor molecule displayed by some T-helper cells. Once attached, the virus enters the cell. The viral RNA and reverse transcriptase are released into the cytoplasm of the invaded cell and a complementary DNA copy of the viral RNA is synthesized. It enters the host cell nucleus and is integrated into the cell's genome (Lowenstine and Lerche 1988).

The chimpanzee (*Pan troglodytes*) and the gibbon (*Hylobates* species) are susceptible to HlV-1 infection and the virus can be isolated from the lymphocytes in their peripheral blood but neither animal develops signs of disease. A number of isolates of HIV-2, the AIDS-inducing virus endemic in human populations in West Africa, have also been administered to and studied in nonhuman primates (baboons [*Papio*] species and a number of macaque species), but again no disease developed from these infected primates. A number of HIV-related viruses of nonhuman primates have been identified. These SIV include SIVmac, first isolated from a rhesus monkey with a lymphoma; SIVagm, first isolated from a healthy African green monkey *Cercopithecus aethiops*; SIVsmm, first isolated from a healthy sooty mangabey monkey *Cercocebus atys*; and SIVmnd, first isolated from a healthy mandrill *Mandrillus sphinx* (Letvin and King 1990). These isolates all share significant nucleotide homology with HIV-1 and HIV-2 as well as tropism for CD4 lymphocytes and monocytes/macrophages. A notable point is that the isolates from the mangabey,

African green monkey, and mandrill (all African species) do not produce disease in these animals and it is believed that the latter animals are natural hosts of the virus. However, SIVmac and SIVsmm and others when administered to rhesus monkeys (and a variety of Asian macaque species) induce a disease with a remarkable similarity to human AIDS (Letvin and King 1990). (One would assume that the lymphomatous rhesus monkey from which SIVmac was isolated was truly infected and was not the "natural host" of the virus that it carried.)

The clinical presentation of SIV disease in the rhesus monkey includes a lymphadenopathy syndrome and immunologic abnormalities with qualitative and quantitative T-cell dysfunction. The monkeys die with opportunistic infections and tumors. The clinical course of the induced simian disease has some notable differences from the HIV-induced infection in humans. While the disease evolves in humans over a period of years to decades following infection, SIV-infected macaques usually die within 2 years. In addition, two rather distinct clinical progressions of the disease are seen. On one hand, perhaps one-third of infected animals develop a very rapid disease that leads to death within a few months of inoculation while the others develop a more slowly progressive disease that terminates in death in 1–2 years.

Macaca mulatta monkeys infected with SIV are prone to demonstrate hematologic abnormalities that are associated with the disease but whose specific relationships with the infection are obscured because of multiple concurrent confounding factors. This picture parallels the complex hemopoietic profile of HIV-infected human patients who display variable manifestations depending on the clinical progress of the infection, the patient's individual response to the virus, and the medical treatment given to the patient (Thiebot et al. 2001). Zon and his coworkers (1987, 1988) have quantitatively analyzed the many clinical manifestations of (human) individuals infected with HIV and this condition's recognized differing levels of disease development.

An investigation has been conducted on unpretreated *Macaca mulatta* that were inoculated with SIV and monitored thereafter in the absence of treatment throughout the course of their infection as it developed into AIDS (Gill et al. 2012). With the avoidance of many confounding factors, an improved insight into the direct specific effects of SIV on the hemic profile of the rhesus monkey was achieved.

The subjects of the previous investigation were 1.8–12.2 years old with a mean age of 4.2 years. The group totaled 26 *Macaca mulatta*; each individual was inoculated with a single dose of pathogenic SIV, SIVmac239, SIVmac251, or SIVB670. This was a retrospective study that was conducted over a period of 20 years. As indicated, the study excluded any animals that had been vaccinated, immunized, and/or challenged with other viruses; treated for SIV or other diseases; administered antiretroviral therapy or other antimicrobials; or were rapid progressors of the disease. Criteria for cohortship also included a normal progression to AIDS (≥ 260 days post-inoculation) and serial complete blood counts with at least one CBC during each successive stage of infection until humane sacrifice. Criteria for the diagnosis of AIDS were based on the 1993 Centers for Disease Control (CDC) guidelines and limited to opportunistic infections, encephalitis and/or multiorgan lymphoma confirmed by histopathologic examination at necropsy. The hematologic data were organized and evaluated according to the stage of infection: that is, uninfected or preinoculation controls, acute stage (1–42 days postinoculation [dpi]), chronic asymptomatic stage (43–120 dpi), and AIDS (>120 dpi). Prevalence rate was defined as the percentage of total macaques with a particular hematologic abnormality, while incidence rate was defined as the percentage of total participants observed with the first event of a hematologic abnormality within a defined period. The different inoculum groups yielded statistically equivalent results. The prevalence rate for anemia was 3% for naïve subjects and 88% for virus-infected macaques (P ≤ 0.05). (The corresponding observations for other hemopoietic cells for naïve versus infected monkeys, respectively, included the following: lymphopenia, naïve 3% versus 100% for infected subjects; eosinophilia 13% versus 94%; neutropenia 0% versus 91%; neutrophilia 10% versus 70%; and thrombocytopenia 3% versus 66% for healthy and infected monkeys, respectively. Lymphopenia was observed as the highest prevalence for a hematologic abnormality during SIV infection and was also cited as a hallmark for all stages of infection. It was present in all inoculum groups with at least 40% prevalence ending with 100% prevalence in the AIDS stage. It resulted in CD4 T-cell depletion.) When the prevalence of anemia in infected monkeys was considered relative to the stage of infection in comparison with its occurrence in control monkeys, both acute and AIDS phases had a significantly higher occurrence of anemia (P ≤ 0.001) while no significant difference was identifiable in the chronic state. Overall, the anemia was mild, normocytic, and normochromic yet highly prevalent in infected animals. Anemia was defined as a low hematocrit value, <34.8%. Reticulocyte counts were not performed and so were not available to ascertain regeneration during anemic episodes. A lack of significant differences in MCV in control and infected animals was noted. Several causes of anemia were ruled out including hemolysis, renal disease, endocrine disease, bone marrow infiltrative disease, and myelodysplasia. Thus, the anemic state was suggestive of a nonregenerative process such as anemia of chronic inflammatory disease or early iron-deficiency anemia. Anemia was often recurrent but rarely persistent in infected subjects.

In summary of the preceding study, virtually all of the previous SIV-infected animals displayed anemia (as well as thrombocytopenia, neutropenia, neutrophilia, lympopenia, or eosinophilia) at some stage of SIV infection. Anemia exhibited its highest prevalence and incidence in the AIDS stage. The analyzed infected *Macaca mulatta* were normal progressors of AIDS devoid of therapeutic intervention. They recapitulated the occurrence of several hematologic abnormalities observed during untreated HIV infection in *Homo sapiens*.

Infection of SIV in macaques (and HIV-1 in humans) causes destruction of CD4+ T cells (one of the hallmarks of progression of the disease). Monocytes/macrophages are also important targets of SIV/HIV infection (Mandell et al. 1992; Braun et al. 2013). The latter workers have further noted that a massive turnover of peripheral monocytes and the demise of tissue macrophages in SIV-infected macaques were not linked with the CD4+ cell count and were a better predictor for AIDS progression than were viral load or lymphocyte activation. In order to investigate the impact of the increased turnover of monocytes on hemopoiesis in SIV-infected *Macaca mulatta*, the bone marrow of eight rhesus monkeys was monitored postinfection by colony-forming unit (CFU) assay and by incorporation of BrdU into CD34+ hemo-poietic stem cells (Braun et al. 2013). (BrdU is bromodeoxyuridine [5-bromo-2′-deoxyuridine], a synthetic nucleoside. It is an analog of thymidine and can be used in the detection of proliferating cells. It is incorporated into newly synthesized DNA and its presence is recognized by immunohistochemical staining.) It was seen that the frequency of CFU-GM (colony-forming unit-granulocyte monocyte, the imme-diate precursor of neutrophils and monocytes) increased from 68% of the population preinfection to 89% 50 days postinfection. This was interpreted as an indication of a corresponding decrease in erythrocytic and megakaryocytic progenitor cells. The bulk of marrow CD34+BrdU+ was not diminished. These data suggest that the hemopoietic stem cell pool may be constant in SIV-infected *Macaca mulatta* but the diversion of hemopoietic activity of CFU-GEMM (the myeloid stem cell) toward CFU-GM and a deficiency in the generation of erythroid and megakaryocytic precursors eventually lead to anemia and thrombocytopenia that is seen prior to the death of SIV-infected monkeys. It would seem that this finding is consistent with the concept that the anemia obtained in SIV disease is a nonregenerative anemia as proposed by others (Gill et al. 2012).

Other investigations suggest that the anemia of SIV-infected *Macaca mulatta* is multifactorial (Hillyer et al. 1993). Bone marrow cultures were obtained from normal control monkeys (n = 6), SIV-infected but "well" (n = 11) and SIV-infected classically end-stage "sick" cohorts (n = 5). "Sick" monkeys' marrows demonstrated a drastic quantitative per plate reduction in the generation of cultured BFU-E in comparison with "well" SIV-diseased monkeys, which in turn also generated fewer BFU-E than normal bone marrow (but more than end-stage participants). At both levels of the disease, the infected animals yielded statistically significantly fewer BFU-E than the healthy controls. Thus, it was seen that the seriously ill subjects as well as the infected healthier participants (but to a lesser degree) manifested a deficiency in BFU-E, the progenitor of subsequent precursors committed to red cell production. A consequent deficiency in erythrocyte production, that is, an anemic state would be the logical outcome particularly if the disease process continued to progress. A simi-lar decrease in bone marrow erythropoietic progenitor cell BFU-E colony formation was identified in SIV-infected (SIVmac) rhesus monkeys by Watanabe et al. (1990). In this group's experience, however, a reduction in CFU-GM (colony-forming unit-granulocyte monocyte) colony formation was contemporaneously noted. The virus in this study was isolatable from macrophages, and conversely, inhibition of SIVmac replication in bone marrow macrophages resulted in increased progenitor cell colony growth from bone marrow cells.

Furthermore, Hillyer et al. (1993) also observed that direct antiglobulin tests were positive in 9 of 35 "well" SIV-infected and 14 of 14 "sick" SIV-diseased mon-keys. None of 25 control subjects had a positive direct antiglobulin test. Animals with a positive direct antiglobulin reaction presented a moderate to severe ane-mia. Polychromasia, anisocytosis, and/or spherocytosis was seen in all "sick" SIV-infected monkeys. The mean Hcts were 40.9%, 38.7%, and 25.6% for normal, "well" SIV-infected, and "sick" SIV-infected monkeys, respectively; the reticulocyte counts for the same groups were 1.5%, 2.1%, and 5.5%, while the MCVs were 75, 73, and 81 fL. Thus, in addition to the deficient generation of BFU-E, the SIV-infected rhesus monkeys concurrently presented evidence of an IgG and complement-mediated autoimmune hemolytic anemia. The mean LDH (lactate dehydrogenase) level was 426 U/L for healthy controls (n = 13), 474 U/L for "well" SIV-infected (n = 21), and 1372 U/L (n = 9) for "sick" SIV-afflicted monkeys. Erythroid hyperplasia was noted as well as dyserythropoiesis in bone marrow aspirates of the "sick" infected group. Three members of the latter cohort developed severe anemic manifestations, that is, Hct < 10%, reticulocyte count >16%, and 13–18 nucleated erythrocytes per 100 wbc on blood smears. The mean erythropoietin level was 4.0 mU/mL in normal controls. It was increased to 15 mU/mL in the "well" infected subset and aboundingly elevated to 1176 mU/mL in the "sick" HIV-infected macaques. An earlier report also described one erythropoietically normal, gravid female *Macaca mulatta* (Hct 40%, MCV 70 fL, normal Rbc cytology) that was experimentally inoculated and infected with SIVsmm9 and subsequently developed a severe autoimmune hemolytic anemia (Hillyer et al. 1991). She was inoculated in April (1989) and gave birth to a female infant in the following July. In April of the following year (1990) on through October this adult female became anemic and displayed moderate anisocytosis, poikilocytoisis, spherocytosis, and reticulocytosis. She also presented with generalized lymphadenopathy, splenomegaly, and weight loss. A direct antiglobulin test was positive. Monospecific reagent testing showed both IgG and complement on the surface of the red cells helping to establish the diagnosis of warm (IgG-type) autoimmune hemolytic anemia. The Hct ultimately declined to 6% at which point she was humanely killed. The infant had a hematocrit of 36% at 6 months of age. She was diagnosed as failing to thrive at one year, was anemic, had a 16% hematocrit, circulating spherocytes, and 13 nucleated erythrocytes per 100 leukocytes on peripheral blood smear. She also had a direct antiglobulin test positive IgG autoimmune hemolytic anemia like her dam. Of 13 other gravid monkeys that comprised the inoculated cohort in this investigation, 4 subjects had weakly positive antiglobulin tests but were not anemic.

Histologic analyses of the bone marrow of SIVsmm-infected *Macaca mulatta* reveal cytological evidence that this condition does affect the erythroid and other hemopoietic cell lines. The greatest alterations are seen in the more seriously sick subjects. An examination of posterior iliac crest Wright–Giemsa-stained smears and hematoxylin- and eosin-stained core biopsies of nine SIVsmm-infected monkeys with clinical findings limited to lymphadenopathy and/or splenomegaly, 12 infected "ill" cohorts that had additional findings of diarrhea, wasting, and/or peripheral blood abnormalities including anemia, and thrombocytopenia along with eight seronegative controls was conducted by Klumpp et al. (1991) (Table 13). Erythroid hyperplasia and dyserythropoiesis, respectively, were diagnosed in 6 and 10 of the seriously sick subjects. Of the healthier but infected ill cohorts, one presented dyserythropoiesis and none displayed erythroid hyperplasia. Pancellular hyperplasia was obtained most often in "sick" subjects (n = 5). It was also seen, but less frequently, in "healthier" SIV infected (n = 1) as well as in controls (n = 2). Other findings included megakaryocytic dysplasia in four and six of relatively healthy and clinically ill monkeys, respectively. This latter presentation is pertinent in view of the fact that the virus

Table 13   Bone Marrow Morphology of Simian Immunodeficiency Virus
(SIVsmm)-Infected *Macaca mulatta*

Subjects	Seronegative Controls	SIV+ with Lymphadenopathy and/or Splenomegaly	SIV+, Sick
Number of subjects	8	9	12
Subjects with pancellular hyperplasia	2	1	5
Subjects with erythroid hyperplasia	—	—	6
Subjects with dyserythropoiesis	—	1	10
Subjects with megakaryocytic dysplasia	—	4	6
Subjects with myelofibrosis	—	2	1
Subjects with lymphoid aggregates	—	—	5
Subjects with granulomas	—	—	1

Source: Klumpp, S.A. et al., Significant bone marrow abnormalities in SIVsmm-infected rhesus macaques: analysis of aspirates and biopsies, in *Proceedings of the Ninth Annual Symposium on Nonhuman Primate Models for AIDS*, Seattle, WA, 1991, p. 105.

Notes: SIV+ subjects manifested lymphadenopathy and/or splenomegaly only. SIV+ sick subjects had additional findings of diarrhea, wasting, and/or peripheral blood abnormalities including anemia and thrombocytopenia. Posterior iliac crest aspirate Wright–Giemsa-stained smears and hematoxylin–eosin-stained proximal femoral core biopsies.

has been shown to invade megakaryocytes. Bone marrow hypercellularity, myeloid, megakaryocytic, and macrophage hyperplasia have been documented in early SIV infections of rhesus monkeys in other studies (Mandell et al. 1992, 1993).

A relatively early study of 31 rhesus monkeys with simian acquired immune deficiency syndrome (SAIDS) was conducted in which 15 subjects had acquired the disease by natural transmission and 16 which became infected by being inoculated with infectious material (blood, plasma, or lymph node homogenates) (MacKenzie et al. 1986). All 15 natural transmission animals died within 6 months of placement in a corral where spontaneous transmission of the disease was occurring. On the other hand, the 16 inoculated animals were followed until the time of death or euthanasia 2–11 months postinoculation (n = 10) or until 1-year postinoculation (animals considered having a chronic disease course, n = 6). The most striking hematologic abnormalities of these two experimental cohorts were the consistent anemia (and a high incidence of granulocytopenia, and in those subjects that survived at least several months, the development of lymphopenia). All but 1 of the 31 test subjects had a hematocrit less than 35% and 23 of these monkeys had an Hct < 30%. Animals that were inoculated all had an initial hematocrit of at least 40%, and as early as 2–3 weeks postinoculation they began to show a decrement of 5–10 hematocrit units. All members of this group exhibited a reduced Hct by 5 weeks postinjection. The monkeys that became infected by natural transmission, conversely, had a slightly slower reduction of their hematocrit. The earliest decrease occurred within 5 weeks of exposure and the majority showed lowered hematocrits by 8–9weeks. A graphic comparison of the lowest hematocrit of each participant in the two groups revealed that the range and distribution of the values were similar in the two groups (MacKenzie et al. 1986). The mean Hct and standard deviation for each cohort were essentially identical (natural transmission: 19.5% ± 9.7, inoculation: 20.3 ± 9.9). The anemia was microcytic and hypochromic; the preexposure and postexposure values, respectively, for the erythrocyte indices were MCV 70 and 64 fL, MCH 23.1 and 19.5 pg, MCHC 33 and 31%. Each reduction was statistically significant. The anemia tended to be progressive in the 27 animals that died of the disease. Most of the reticulocyte counts were around 1%, and calculation of absolute reticulocyte counts was consistent with ineffective erythropoiesis. Solid evidence for hemolysis was absent although serum lactate dehydrogenase was moderately elevated (between 500 and 600 units, as opposed to the normal value of 100–200 units). This higher level was obtained primarily in animals that did not have a progressive anemia. Direct Coombs tests were uniformly negative in animals with the most severe and progressive anemias.

Humeral bone marrow aspirates were obtained from 17 subjects either during life or at necropsy. Fifteen of them presented erythroid hyperplasia, and the maturation was frequently morphologically abnormal. The cytoplasm of the atypical erythroid precursors had an increased basophilia while the nuclei were irregularly shaped. The presence or absence of stainable iron was determined in six samples by the application of the Prussian blue reaction. Five of the specimens yielded a negative reaction.

The erythroid (and leukocytic) abnormalities of both groups of infected monkeys were virtually identical. The cause of the profound anemia in these macaques was interpreted as multifactorial and superimposed with iron deficiency. The latter condition was considered as nutritional and viewed as akin to the iron deficiency observed in other monkeys housed in outdoor corrals (Anderson et al. 1983). Unlike some other investigated SIV-infected rhesus monkeys, there was no evidence of a definite hemolytic anemia. The absence of ring sideroblasts in the bone marrow and concomitant paucity of sequestered iron in the phagocytic cells was proposed as an indication that a major disruption iron metabolism was not in evidence.

Anemia is one of the verified components of SIV infections. The pathogenesis of the anemia, however, is indeed not straightforward. Depending on potential circumstances such as the source of the virus, duration of the infection, and the stage of the condition, the specifics of an investigation including the methodology, varied dynamics may have a role in the deficiency of erythrocytes seen in this disease. Further investigation is required to reach a better understanding of the origin of this anemia. O'Sullivan et al. (1996) have speculated that it is possible for a given monkey with a severe anemia ascribed to its documented infection with a simian retrovirus to have an unrecognized contemporaneous infection with simian retrovirus and this latter organism's copresence could be a major factor for the severity of the anemia.

In an investigation of SIV infection of neonatal monkeys, seven 3-day-old *Macaca mulatta* were inoculated with the simian immune deficiency virus SIV/ DeltaB70 and became infected with the agent (Bohm et al. 1993). Five subjects died within a mean 31 days (identified as short-term survivors). The two others died at

216 and 423 days (long-term survivors). It was noted that the neonatally SIV-infected rhesus monkeys exhibited strong similarities to human pediatric HIV infection. The most obvious correspondence was the rapid progression of disease and early death. A cytological parallel was the absence of lymphopenia. The basis for the distinctly different pictures of adult and infant SIV infections was not ascertained. While the blood picture of the monkeys was monitored with weekly CBCs, an erythrocytic response to the infectious process was not identifiable. Some variation in the immunologic identity of circulating lymphocytes was recognizable. It was hypothesized that a decrease in the CD4+CD29+ subset within 3 weeks postinoculation of the virus may be a prognostic indicator of rapid disease progression and death in the SIV-inoculated neonate.

CHAPTER 20

Erythroid Profile of Pregnant
Macaca mulatta

It has been established that during pregnancy the body undergoes many physiologic adjustments. This would be foreseen even in the absence of investigative data. One of the apparently earliest studies of the blood picture of the rhesus monkey throughout its mean ~165–166-day-long gestation period was carried out by Allen and Siegfried (1966) (Table 1). Forty-two pregnant monkeys were monitored; one aborted on the 89th day of pregnancy and nine subjects failed to deliver naturally at term (of these, three viable and five dead infants were obtained at cesarean section while the remaining monkey delivered a stillborn infant on the 189th day of gestation). Eleven scheduled femoral blood samplings, including an initial prepregnancy phlebotomy, were obtained from all subjects throughout the 179 days of monitoring. The most notable observation regarding the subjects' erythrocytes, in the opinion of the investigators, was the marked elevation in the erythrocyte sedimentation rate (Esr) that increased from an initial mean 1.4 mm/hr (per Wintrobe tube) at around the 120th day of pregnancy to a maximum of 22 mm/hr by the time of delivery. The sedimentation rate declined soon after parturition and returned back to normal by 15–19 days postdelivery (Figure 35). The increased sedimentation rate was attributed to the increase in plasma fibrinogen and globulin. The Esr was used to estimate the stage of pregnancy, time of delivery, necessity for cesarean section, and to forecast possible abortion. It was noted that if there were a rapid increase in the sedimentation rate during early gestation, one would anticipate an abortion. As the monkey approached the time of delivery, little difficulty would be encountered in the gestation as long as the sedimentation rate remained elevated. If a sharp reduction in Esr did present without delivery, the death of the infant *in utero* would occur within 48 hours unless the infant were surgically extracted. The investigators also noted that there was "an absence of any appreciable change in the hematocrit and hemoglobin level" during gestation (Table 1). The mean hematocrit and hemoglobin concentrations for the first approximate 100 days following conception (including the pregravid value) were minimally higher than for the remainder of the pregnancy (Hct and Hb for the first ~100 days versus the remaining ~65 days, respectively: Hct 40.6 and 39.5%; Hb 12.0 and 11.6 g/dL). The data were not statistically analyzed, and erythrocyte counts were not performed.

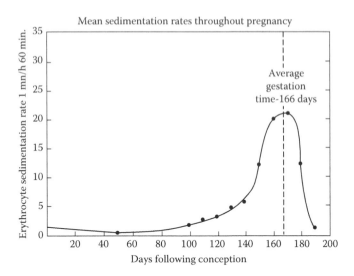

Figure 35 *Macaca mulatta*. Erythrocyte sedimentation rate of *Macaca mulatta* during pregnancy. A marked increase is noted during the interval initiating at approximately day 100 of gestation (mean 1.4 mm/hour) and attaining a maximum (mean 22 mm/hour) at delivery. A rapid decline and return of the basal level occurs soon after delivery (per Wintrobe erythrocyte sedimentation tube). (From Allen, J.R. and Siegfried, L.M., *Lab. Anim. Care*, 16, 465, 1966.)

Another study of the hematologic alterations that accompany pregnancy in the rhesus monkey was reported by the senior investigator of the study discussed earlier (Allen and Ahlgren 1968). Twenty-five subjects were monitored throughout pregnancy. The results were in accord with the initial findings as well as adding further information. The erythrocyte count (determined by electronic counting) was followed throughout the course of the pregnancy and was seen to gradually diminish throughout this period (Table 1, Figure 36). It began at 5.57×10^6 Rbc/μL in the prepregnant state and progressively declined to 4.4×10^6 cells/μL by delivery. As in the earlier study, a modest decrease in the hemoglobin concentration and hematocrit was identified during gestation (Table 1, Figure 36). During the postpartum period, the mean level of each of these major erythroid parameters (Rbc, Hct, Hb) slowly returned toward the original value but did not reach the preconception level by 40 days following delivery. The Esr duplicated the curve observed in the earlier investigation (i.e., a marked increase initiating at ~120 days of gestation). The flows of the erythrocyte count, hematocrit, and hemoglobin concentration of the blood in the rhesus monkey throughout its ~165 days long pregnancy are graphically depicted and compared with the homologous curves for the human during its 33+ weeks long pregnancy (Allen and Ahlgren 1968, Figure 36). It is seen that the pregnancy long curves for the Rbc count and Hb level for the two species are comparable. The hematocrit curves display decreases in the human and the monkey that entail the entire gravid interval but the shapes of their individual curves are not parallel. It should be noted that although the levels of the three major erythroid constituencies (Rbc count,

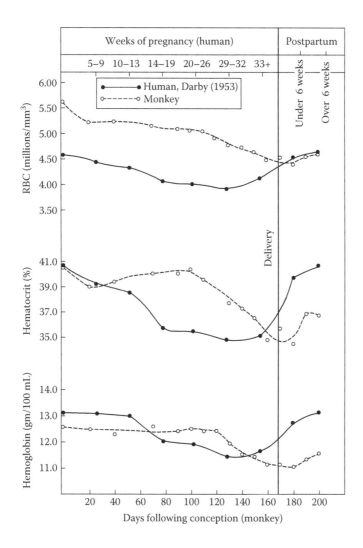

Figure 36 *Macaca mulatta.* Comparisons of the alterations of the erythrocyte counts, hematocrits, and hemoglobin concentrations of the blood during pregnancy of *Macaca mulatta* and the human female. The time lapse of pregnancy in days for the monkey is presented at the base of the graph. The length of gestation for *Macaca mulatta* is ~165 days. The time frames of human pregnancy are given in weeks (5–9 weeks to 33+ weeks) at the top of the graph. When the relative segments of the gestational periods for the two species are compared, it can be seen that the RBCs and hemoglobin curves have very similar forms. (From Allen, J.R. and Ahlgren, S.G., *Am. J. Obst. Gync.*, 100, 894, 1968.)

Hct, Hb) all diminish per unit volume of blood during pregnancy, this does not necessarily indicate an absolute decrement in their total body values because pregnancy
is accompanied by an increase in blood volume by approximately 30% in both the
rhesus monkey and human (Allen and Ahlgren 1968).

The MCV increased from 73 to 80 fL during gestation (Allen and Ahlgren 1968,
Table 1). This would suggest the possibility of increased erythropoiesis and the
enhanced release of reticulocytes (which have larger dimensions than mature red
cells) into circulation. This picture would be consistent with erythrocytic hyperplasia
secondary to the enlargement of the total blood volume during pregnancy and the
requirement for supplementary red cells to be added to the total body pool of circulating erythrocytes.

The profiles of the erythrocyte-dependent values' curves presented during pregnancy in *Macaca mulatta* offer a general overview of erythropoiesis during this
period (Allen and Siegfried 1966; Allen and Ahlgren 1968). The data, however, are
not necessarily very precise because the phlebotomies were conducted under less-
than-ideal circumstances (manual restraint of the subjects, and hence during presumed stress) and the determination of the hematocrits by centrifugation of capillary
tubes containing blood as opposed to the more exact determination by electronic
means. The derived value for the centrifuged hematocrit is also used in the Wintrobe
formula for the calculation of the MCV thereby inducing some deviation from the
ideal calculation of this erythrocyte index.

An assessment of "recently captive" rhesus monkeys (n = 14 mature nonpregnant
females, n = 9–13 subjects in estimated early pregnancy ≤90 days, and n = 4–12 in estimated late pregnancy >90 days) indicated that among these monkeys the hemoglobin
level was slightly but statistically significantly decreased in late pregnancy while the
MCV was significantly increased in this interval (Spicer and Oxnard 1967). These
two observations are consistent with those of Allen and Ahlgren (1968). Spicer and
Oxnard (1967) also stated that they did not observe any significant changes in the
erythrogram during estimated early pregnancy (≤90 days). Other than these general
statements, further conclusions are limited since the studied monkeys were recent
captives and not standard research, long-term laboratory-maintained animals, and
some of the analytical techniques were not contemporary. Hemoglobin was determined as oxyhemoglobin per colorimeter, and the hematocrit was determined by
microhematocrit centrifugation.

A hemocellular (and biochemical) analysis of 97 sexually mature, nonpregnant,
and 97 pregnant 3–10+ year-old rhesus monkeys was conducted by Buchl and Howard
(1997) at the University of Texas, Department of Veterinary Sciences (Table 1). Blood
was drawn from the femoral vein under ketamine anesthesia during an animal's annual
physical examination. The analyses were performed using contemporary, automated
techniques. The elapsed time into pregnancy (the duration postconception) during
which the blood sample was drawn for a given animal was not given. The weighted
mean values for the two groups were very similar. The erythrogram for the nonpregnant monkeys was Rbc $5.79 \times 10^6/\mu L$, Hct 40.3%, Hb 13.0, and MCV 70 fL while the
corresponding data for the pregnant monkeys were Rbc $5.83 \times 10^6/\mu L$, Hct 40.6%,
Hb 13.1, and MCV 70 fL. A recent comparison of 14 pregnant *Macaca mulatta* that

were sampled once between the 6th and 10th week of pregnancy (day 56 ± 14) and an equal number of nonpregnant monkeys failed to demonstrate any statistically significant differences in the mean erythrocyte count, hematocrit, and hemoglobin concentration of the two groups (Ibanez-Contreras et al. 2013, Table 1). This observation could be interpreted as consistent with the consensus that the greatest erythrocellular differences between gravid and nongravid monkeys are seen in the later rather than the earlier segments of gravidity (as specifically illustrated by Allen and Ahlgren 1968, Figure 36). (A significant increase in the circulating leukocyte count and in the number of circulating neutrophils was observed in the pregnant monkeys.) The subjects were captive housed, 4–6 years old. They had an 8-hour fasting period and were administered ketamine and xylazine prior to venisection.

Inasmuch as pregnancy is associated with modifications of the cellular and biochemical picture of the blood, a return to the nongravid profile would be anticipated following delivery. This recovery is one of the natural topics of interest in hematology. An analysis of the postpartum morphologic blood picture of the rhesus monkey was conducted by Switzer et al. (1970). Two experiments were undertaken. The initial one involved 20 monkeys that were analyzed at 24, 48, 72, 96 hours and at 1, 2, 3, and 8 weeks postpartum (Table 1). Complete erythrograms as well as reticulocyte counts were performed. In the subsequent study, 50–200 gravid subjects were followed from the 17th to 53rd day of pregnancy onward through delivery until 46–77 days postpartum. The hematocrit and Esr (as well the total leukocyte and differential counts) were monitored in this larger group of animals. The mean value of the (capillary centrifuged) hematocrit at 24 hours postpartum in the first group was 37.3% and it decreased over the following 2 days (72 hr postdelivery) to 34.4%. Thereupon from approximately 96 hours onward, it gradually increased to 42.4% at 8 weeks postpartum (normal nongravid control Hct = 41.8%). The erythrocyte count was $4.85 \times 10^6/\mu L$ 24 hr after delivery and diminished to $4.68 \times 10^6/\mu L$ during the 96 hr–1 wk interval. It then increased to the "normal" level of $5.78 \times 10^6/\mu L$ by the eighth week (control = $5.84 \times 10^6/\mu L$). The hemoglobin level presented a similar pattern. Its 24-hour postdelivery level (11.5 g/dL) diminished through 96 hours (4 days) to 10.7 g/dL and thereafter increased to 13.4 g/dL by the eighth week postpartum, the normal control level. The average hematocrit of the larger group displayed an early-in-pregnancy (initial ± 20 days) slight recession followed by a modest incline (within the range of normal) that was also illustrated by Allen and Ahlgren (1968, Figure 36). This latter interval of increment, in the experience of both groups of workers, persisted up to ~100 days of gestation and then began a decline that continued to parturition. At delivery, the nonweighted approximate mean of the Hct derived by Allen and Siegfried (1966), Allen and Ahlgren (1968), and Switzer et al. (1970) was 37%. The packed cell volume returned to or toward the typical level at around 25 days after delivery. A comparison of the pregnancy/postpregnancy hematocrit curves presented by Allen and Ahlgren (Figure 36) and Switzer et al. (Figure 37) reveals the similarity of the observations derived from their respective investigations. Not surprisingly, Switzer et al.'s (1970) pregnant macaques confirmed the occurrence of an elevated erythrocyte sedimentation rate that appeared around day 120 of pregnancy.

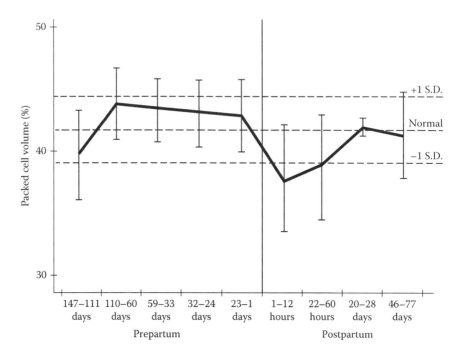

Figure 37   *Macaca mulatta.* Changes in the hematocrit of *Macaca mulatta* during pregnancy and approximately 70 days following delivery. The hematocrit is seen to exhibit a tendency to increase slightly during days ~15–100 of gestation (47–60 days prior to delivery), and then progressively decreases until delivery. This curve is in general agreement with that given by Allen and Ahlgren (1968) in the previous figure (Figure 36). Recovery initiates 1–2 days after delivery and returns to normal levels within 21 days. Vertical bars indicate ± 1 SD. Dotted lines indicate ± 1 SD of the normal level. The prominent vertical line that is almost in the middle of the graph indicates the time of delivery. Values at the base of the graph, left of this division, indicate the number of days prior to delivery. The values to the right of this demarcation indicate the hours or days postpartum. There were 20 subjects that were assayed at monthly intervals during pregnancy and for a period of 8 weeks after delivery. An additional 100–200 animals were hematologically evaluated during gestation on a trimester basis and for 8 weeks postpartum. (From Switzer, J.W. et al., *Lab. Anim. Care*, 20, 930, 1970.)

A brief overview of the status of the major erythroid parameters (Rbc, Hct, Hb) in *Macaca mulatta* during its ~165 day gestation and following delivery indicates that the levels of these quanta tend to decrease during pregnancy particularly after 100 days postconception. Each of these values thereupon decreases from the 24-hour postdelivery level until approximately 96 hours to 1 week postpartum. From this juncture onward, the values increase until the normal levels are attained at 4–8 weeks postdelivery.

The number of reticulocytes in the circulating blood during pregnancy and postpartum might be expected to deviate from the usual level. The concentration of reticulocytes in the blood of a normal, adult nongravid rhesus monkey is roughly 1% of its erythrocyte count (equivalent to the content observed in a normal adult human).

Switzer et al. (1970) reported that their normal monkeys (n = 77) had a mean reticulocyte count of 0.7% and an erythrocyte count of 5.84 × 10⁶/μL (absolute reticulocyte count 40,880/μL). The rough mean absolute reticulocyte count in the monkey (and man) is frequently cited as 50,000/μL of blood. At 24 hr postpartum, the reticulocyte count is an elevated 121,250/μL blood, 2.5% of a reduced (from normal) red cell count of 4.85 × 10⁶/μL (n = 13) (Switzer et al. 1970). The absolute reticulocyte count continues to increase until 1 week postdelivery at which point the absolute count is 257,400/μL blood (5.5% of 4.68 × 10⁶/μL, n = 18). At 2 weeks postpartum, the absolute count begins to decline (174,240/μL, 3.3% of 5.28 × 10⁶/μL). It continues to be reduced to the probable normal range of 73,406/μL (1.27% × the normal red cell level of 5.78 × 10⁶/μL, n = 18) at 8 weeks after delivery (Switzer et al. 1970). The latter point in time is when the erythrocyte count, hematocrit, and hemoglobin have also regained their standard levels. All of the physiologic reasons for an enhanced reticulocyte count following delivery are not apparent.

The mean serum concentration of iron of mature female nonpregnant *Macaca mulatta* was reported by Spicer and Oxnard (1967) to be 171 μg/dL (n = 9). It did not change significantly in the first 3 months of pregnancy but was significantly lower in the last 3 months (120 μg/dL, n = 7). A corresponding low serum iron has been found in late human pregnancy.

CHAPTER 21

Erythroid Profile of Pregnant
Macaca fascicularis

The blood picture of pregnant *Macaca fascicularis* (n = 14) has been followed from the time of mating until the eighth postpartum week by Fujiwara et al. (1974). The levels of the major erythroid parameters (Rbc, Hct, Hb) were stable and essentially unchanged from the control levels until the last 4 weeks of pregnancy when a modest anemia presented, continued to develop, and persisted through the second postpartum week (Table 1). At this point, a recovery was initiated and normal pre-pregancy values were attained at the fourth postpartum week. The more recent investigation of Yoshida et al. (1988) also established in these investigators' experience with 142 pregnant cynomolgus monkeys that test subjects, which were in the last of four segments of pregnancy (i.e., >121 days), similarly displayed diminished levels of the erythrocyte count, hematocrit, and hemoglobin concentration of the blood. The length of gestation of 26 pregnant *Macaca fascicularis* followed by Fujiwara and coworkers (1974) was on average 164 days (s.d. ±7) with a range of 150–170 days. The durations of gestation for the rhesus and cynomolgus monkeys are thereby the same. The long-tailed macaque demonstrated, as did the rhesus monkey, a marked increase in its erythrocyte sedimentation rate during the last 4 weeks of gestation. The rate began to decrease (i.e., initiate a return toward the normal level) at ~1 week prior to parturition and continued to recede during the delivery and postdelivery interval (Fujiwara et al. 1974). The sedimentation rate returned to the nonpregnant status by the second postpartum week.

CHAPTER 22

Erythrocytes and the Menstrual Cycle

The macaques have a menstrual cycle just like the humans. The erythrocyte count has been shown to vary with the phase of the cycle. The first day of a given cycle is the first day that the monkey menstruates. The cycle persists until the first day of the next episode of uterine hemorrhage. The length of an individual cycle varies among individuals and may roughly range from 26 to 30 days in the rhesus monkey (Fortman et al. 2002). The first half of the menstrual cycle is termed the follicular period and ovulation occurs at the end of the first half of the cycle.

Some of the initial analyses of the relationships of the menstrual cycle to the erythrocyte count in the rhesus monkey (and also *Macaca nemestrina* the pig-tailed monkey) were conducted by Guthkelch and Zuckerman (1937). They observed that the erythrocyte count rose rapidly during the first half (follicular phase) of the menstrual cycle, then dropped to a low at ovulation, and subsequently rose slightly before menstruation. For example, in one rhesus monkey the red cell count increased from 5.18 to 5.73 million per microliter in the first segment of the cycle, then receded to 4.59 million, and thereupon increased slightly before menstruation. *Macaca nemestrina* presented the same pattern.

Ovariectomized mature rhesus monkeys (n = 5) were given daily injections of oestrone (synonym: estrone, an estrogen) for 14 days. Erythrocyte counts were performed twice each week, and the subjects were observed for uterine bleeding to indicate the onset of the following cycle at which time the regimen of hormone administration was repeated (Guthkelch and Zuckerman 1937). Two cycles were monitored with the second program of injections starting on the second day of bleeding. Consistent with the pattern observed in intact healthy mature macaques undergoing the follicular phase of the menstrual cycle, the oestrone-injected ovariectomized subjects demonstrated an increase in the Rbc count during the course of the injections. After cessation of the administration of the hormone, the red cell count gradually fell during the period that intervened between the cessation of injections and the onset of uterine bleeding. A second increase in the red cell count was observed before uterine bleeding in six of nine experiments. The erythrocyte counts during the investigation (n = 5, 9 cycles of injections) were the following. The mean erythrocyte count during the first week of injections of a given cycle was $4.86 \times 10^6/\mu L$; the mean maximum during the second week of injections was $5.67 \times 10^6/\mu L$. The average minimum count during the latent period prior to bleeding was $4.83 \times 10^6/\mu L$, while the mean red cell count at the onset of bleeding was $4.98 \times 10^6/\mu L$. The studies of intact normal monkeys demonstrating menstrual cycles and oestrone-injected ovariectomized counterparts demonstrated that the erythrocyte count is lowest at the beginning and end of the menstrual cycle. It is highest in the middle of the cycle. The fall in the red cell count coincides with the time of ovulation. The variation of the red cell count is believed to be due to hemo-concentration secondary to hormonal influence. The latter concept is supported by comparable investigations in the pig-tailed monkey.

CHAPTER **23**

Reticulocytes

Reticulocytes are denucleated but nevertheless "not-quite-mature" erythrocytes that are morphologically identified by virtue of their cytoplasmic residual ribonucleoprotein that is typically microscopically visualized by supravital staining with stains such as new methylene blue or brilliant cresyl blue. Electronic counting methods that can be expected to be both less time consuming and more accurate have also been developed (e.g., laser flow cytometric quantitation with a stain that binds to RNA). The material (ribosomes when observed under transmission electron microscopy) is seen under light microscopy as a fine granular-like precipitate that is randomly scattered throughout the cell or localized in one area. Avians and poikilotherms, (which unlike mammals do not extrude the nucleus from their erythrocytes) also have reticulocytes that are likewise identified on the basis of cytoplasmic ribonucleoprotein. In this instance, however, the nucleus is, for example, permanently retained in the cell along, in many cases, with a minimal amount of ribonucleoprotein. Thus, in this latter type of circumstance, morphologic guidelines have to be established to determine whether or not a given erythrocyte that presents some supravitally stained granules should be classified as a reticulocyte or as a mature red cell.

The maturation of the erythrocyte, as discussed previously in the segment Bone Marrow: Erythropoiesis, eventually attains a semi-ultimate stage during which it extrudes its nucleus. It is likely to still maintain a slightly less red, grayish color (in comparison with mature Rbc) when stained with a Romanowsky-type dye such as Wright stain. Such cells are classified as polychromatic (polychromatophilic) erythrocytes. The basis for the homogeneous diffuse off-red staining is the persistence of some RNA (and as much of this material is lost from the cell as it matures, the property of staining this special reddish tint is then lost from the cell). The newly denucleated red cells remain in the bone marrow for a short period (2 days in man) and lose enough of the residual ribosomes to acquire the normal so-called brick red color of mature erythrocytes. These recently denucleated cells, nevertheless, when supravitally stained with a suitable stain will demonstrate stained granular material or dots identifying them as reticulocytes. It is understood that a polychromatophilic erythrocyte whether it is in the bone marrow or in the blood will be stained by the supravital technique and thereby identify itself as a reticulocyte. Of course, such cells are already identified by their somewhat special staining with Romanowsky dye mixtures as immature cells and consequently do not need to be identified as immature by reticulocyte staining.

Most investigators are in agreement that under normal healthy conditions poly-chromatophilic erythrocytes are not released into the circulation of the rhesus monkey (or man). Not surprisingly, occasional citations of minimal polychromato-philia have been made for this nonhuman primate (Krumbhaar and Musser 1921).

On the other hand, erythrocytes that have just been just released from the bone marrow still retain a minimal amount of ribonucleoprotein but have the appear-ance of older cells in circulation that no longer have this cytoplasmic component. The residual ribonucleoprotein is lost from the cell during the first day it is in the circulating blood in man and presumably during the same length of time in the rhesus monkey. The "disappeared" (i.e., matured) reticulocytes are incorporated into the pool of mature red cells in the blood and are replaced by the reticulocytes that are daily released from the bone marrow. Under normal, steady-state condi-tions, the pool of reticulocytes in circulation is stable and constant. A change in the normal proportion is an indication of absolute or relative erythropoietic hyperplasia or hypoplasia.

Polychromatic erythrocytes can frequently be recognized in Wright-stained marrow films (and under certain conditions in blood smears) as being slightly larger than normal adult brick red erythrocytes. Normally, stained brick-red erythrocytes that are reticulocytes (which can be shown to belong to this latter population by reticulocyte staining) are probably also slightly macrocytic. This enhancement in size is probably an accompaniment of immaturity. The polychromatophilic red cells are also often irregularly shaped. This is due to the fact that the terminal normoblast undergoes cytoplasmic contortions in the mechanics of nuclear extrusion. These gyrations induce cellular asymmetry and the contortions continue for a short period even after the loss of the nucleus. This motion during and after nuclear extrusion has been microcinephotomicrographically documented by Bessis (1960) and is respon-sible for the asymmetric polychromatic erythrocytes seen in stained dry film smears obtained in some conditions with enhanced erythropoiesis and increased release of immature red cells into circulation.

The quantity of reticulocytes in a given sample of blood has been usu-ally reported as their percentage of the erythrocytes in circulation. The size of the population in a normal subject is approximately 1% of the erythrocyte count. Alternatively, the reticulocyte count can be expressed as an absolute numerical value, that is, the actual number of these young cells in a unit volume of blood. This offers a more insightful estimation of the reticulocytes' representation in the circulation. The absolute reticulocyte count in the normal rhesus monkey (or man) is an estimated 50,000 reticulocytes/μL of blood (1% × 5.0 × 10^6 erythrocytes/μL blood = 50,000). The usefulness of the absolute count becomes apparent in a theo-retical situation in which a reticulocyte count of 2.3% is derived from an anemic monkey. This might be assessed as a modestly elevated value. Assuming that the erythrocyte count was 2.1 million/μL it would be seen that the 2.3% yields only 48,300 reticulocytes/μL. This level is insufficient to cover the normal daily loss of aged red cells (i.e., less than the estimated required 50,000 cells/μL) and does not aid in the replacement of the deficit of erythrocytes. Hence, this elevated percentage fails to indicate that the reticulocyte count is not genuinely increased.

Accelerated erythropoiesis is believed to foster a skipping of the final mitosis in the maturation of a given erythroblast. This is interpreted as an attempt of an individual's erythropoietic system to immediately increase the number of circulating erythrocytes in the blood during a period of need such as following a significant hemorrhage or during a hemolytic anemia. It occurs in tandem with erythroblastic hyperplasia. This mechanism is also rationally conjectured to occur in the rhesus monkey. The occurrence of this compensatory activity is believed to have subsequent manifestations. The erythrocytes that are the result of a skipped mitosis will be macrocytic and polychromatophilic. It is further envisioned that these cells spend less time than usual in the marrow as polychromatophilic erythrocytes prior to their release into the circulation. The length of time they are in the marrow is theorized to be 1 day (instead of the customary 2 days). Consequently, such "early released" red cells, it is believed, will require 2 days (instead of one) to complete the normal conversion from a reticulocyte to a normal mature erythrocyte. That is, these immature erythrocytes will spend 2 days in the circulation as identifiable reticulocytes rather than 1 day. Such prematurely released reticulocytes have been called stress or shift reticulocytes. One of the aspects of this phenomenon is that the significance of the reticulocyte count is altered. This in effect doubling of the reticulocyte count (because of the longer persistence of reticulocytes) necessitates a change in their physiologic quantitation. For example, the evaluation of whether a subject's reticulocyte response to a need for enhanced erythropoiesis for a given challenge is adequate requires recognition of this longer duration of circulation in the blood. Simple mathematical formulas have been developed to assist a hematologist's evaluation to determine whether, in light of a human subject's hematocrit level, a given reticulocyte count is adequate to address a current erythrocytic deficit. It is likely that the formulas are also applicable to the macaque's hemopoiesis because of the similarity of monkey and human erythropoietic kinetics. The erythrocyte count, hematocrit, erythrocyte life span, and normal reticulocyte count of both groups are reasonably similar, and consequently, the estimation of the reticulocyte response to an anemic condition should be appraisable by the same mathematical formula. In one representative evaluation, a Reticulocyte Production Index is determined in two steps. First, the corrected reticulocyte count is established by the following formula: (reticulocyte count of the subject in % × the subject's Hct in % divided by the average normal hematocrit value). Second, the derived corrected reticulocyte count in % is divided by the maturation duration of the reticulocytes of the subject under evaluation. The maturation duration of the reticulocytes presently in circulation is an estimation of the length of time that these cells will circulate before they mature into normal adult red cells. It is assumed that the greater the deficit of erythrocytes (i.e., the lower the hematocrit) the less mature will be the red cells that are being released from the marrow. Thus, when the Hct level is 36%–45% the anemic stress is probably minimal, and hence this status is assigned a normal maturation factor of 1. However, if the current Hct is in the range of 20%–25%, it is believed that the marrow will release its red cells at a definitely younger age and that these cells will consequently persist as reticulocytes for a longer than normal duration (2 days, a maturation factor of 2). Maturation factors are thus based on the level of the hematocrit and range from 1 to 2.5. A Reticulocyte Production Index equal to or greater than three is considered to represent an adequate response to a given anemic state.

The earliest recognition of reticulocytes in the blood of the rhesus monkey (*Macacus rhesus*, in former terminology) may have been by Krumbhaar and Musser (1921). Nine healthy monkeys were studied. Their blood was supravitally stained with brilliant cresyl blue. Positively stained cells, that is, reticulocytes were described as very rare, occasionally not identified in a given subject, and never present in excess of 0.3% of the red cell population that averaged 5,060,000/mm^3. The distributional patterns of the precipitate were described as skeins (i.e., resembling lengths of loosely coiled and knotted thread), a term that continues to find occasional application in reticulocytic morphology. The skeins were described as much fainter and more delicate than those found in human and canine erythrocytes.

The quantitative occurrence of reticulocytes in the blood of laboratory-housed, singly caged rhesus monkeys (6–7 years old) was investigated in extended, straightforward and illuminating experiments conducted by Harne et al. (1945). The test subjects were schooled over a period of several months prior to experimentation to leave the cage and climb upon a nearby table. They were covered with a stabilized net, petted, and offered favorite articles of food. Eventually, they became accustomed to the procedure, accepted the food, and accommodated themselves to the handling. Blood samples for reticulocyte estimations were taken from the ear by quickly snipping the free border of the auricle with scissors and collecting a small amount into a pipette containing a solution of brilliant cresyl blue stain. In one experiment, a 7-year-old female and a male of the same age were subjected to daily reticulocyte counts for 185 and 174 days, respectively (Figure 38). The female presented a mean reticulocyte count of 10.5/1000 erythrocytes while the male yielded a mean of 10.9/1000 red cells. Assuming theoretical erythrocyte counts of 5 million Rbc/μL for each subject, it is seen that the mean reticulocyte counts for these two monkeys would be ~53,500 reticulocytes/μL (i.e., equal to the approximate value of 50,000 reticulocytes/μL of blood that is often cited as the rough normal absolute reticulocyte count for rhesus monkeys). In another segment of Harne et al.'s (1945) studies, nine other monkeys were subjected to daily reticulocyte counts for periods of 20–81 days. Their mean reticulocyte representation was the same, 10.9 reticulocytes/1000 erythrocytes. The highest average level for an individual animal was 16.7/1000 erythrocytes, which was determined over a period of 20 days of longitudinal study. The same monkey yielded a value of 8.8 reticulocytes/1000 red cells over another 20 days of daily sampling. Although the initial value seemed high, subsequent analysis indicated that the value was correct (both series were introductory to subsequent hemorrhage and determination of the erythrocyte life span that were found to be virtually identical, that is, 105 and 103 days). The lowest value was 8.5/1000 Rbc that was derived from two subjects that had been monitored for 70 continuous daily testings. It was summarized that the daily production of reticulocytes for individual rhesus monkeys showed considerable variation but that over the long term the generation would yield the general mean level observed for the species. This long-term daily fluctuation around the mean is illustrated in Figure 38 and in the stable post-phlebotomy segment of reticulocyte production study of a previously phlebotomized subject depicted in Figure 22. Harne et al. (1945) have suggested that this normal variability of reticulocyte counts prevents the recognition of short-term averages as definitive reticulocyte counts. This group of investigators also studied the effect of defined hemorrhage upon the level of reticulocytes in circulation. Healthy adult monkeys of both sexes underwent femoral vein phlebotomy of a volume of blood equivalent to 1% of their body weight (~40 mL, n = 7) or 0.5% on four occasions at intervals of 72 hr (n = 4). The reticulocyte response for both groups was typically equivalent. It had three phases (as also cited elsewhere). The first extended over a period of 20–30 days during which the reticulocytes rapidly increased in number and subsided to the control level or lower. In the second or interval phase, the reticulocyte count remained at approximately the control level that is followed by spontaneous increase in these cells (third phase). This latter phase was considered to be a reticulocyte response due to the age-related *en masse* loss of the red cells that had been released into the blood as reticulocytes following the bleeding. As noted, it was seen that both types of phlebotomies generated reticulocyte responses of similar magnitude. The peak levels following phlebotomy were in the range of 25–50 reticulocytes/1000 erythrocytes. A cursory examination of this enhancement would suggest (taking into account the probable accelerated erythropoiesis, the resultant earlier release of the reticulocytes from the bone marrow and subsequent doubling of the length of their recognition in the blood to 2 days) that hemorrhage of a volume of blood equivalent to 1% of the body weight in a rhesus monkey results in the increase of the level of erythropoiesis to two or three times normal. It is also interesting to consider at this opportunity the fact that the volume of blood of the rhesus monkey is often estimated to be in the range of 54 mL/kg of body weight and since the aforementioned monkeys weighed ~4 kg, their total blood volume was in the range of 215 mL. Seeing as they were bled ~40 mL, this quantity represented about 18%–20% of their total blood volume.

Reticulocyte counts of young laboratory-housed 10–15 months old *Macaca mulatta* were determined to be a mean 0.87% of circulating Rbc with an average absolute count of 56,600 reticulocytes/μL of blood for males (n = 20) and a mean 0.85% of the red cell population coupled with a mean absolute count of 51,800/μL of blood for females (n = 13) (Winkle 1951). The erythrocyte counts were performed manually using a Levy-Hausser counting chamber (Table 1).

The hemograms of rhesus monkeys of various ages and both sexes were regularly monitored over rather extended periods by Stanley and Cramer (1968, Table 1). The surveys included reticulocyte counts. The test subjects were wild-caught Indian *Macaca mulatta* and consisted of 79 males and 62 females of 2–6 years old. The monkeys were given a total 72 days of postcapture conditioning prior to experimentation. They were housed singly. One group of five male and six female adults (age 3½–6 years) was sampled by 5 mL femoral venisections once per week prior to the morning feeding for 40 weeks. The mean reticulocyte count (total 198 observations) for the males was 0.91%, (s.d. ± 0.39) (mean erythrocyte count 5.54 × 10⁶/μL). The absolute reticulocyte count was 50,400/μL. The corresponding values for the females (230 observations) were a mean reticulocyte count of 0.87% (s.d. ± 0.35)

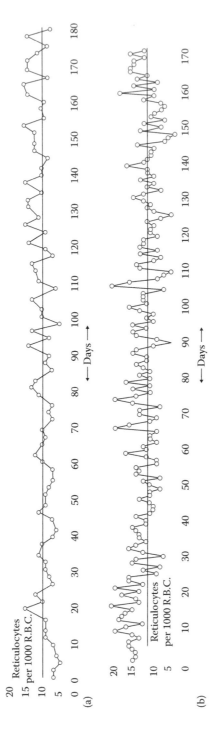

Figure 38 *Macaca mulatta.* Reticulocyte counts monitored on a daily, longitudinal basis over periods of 185 and 174 days (expressed as the number of reticulocytes per 1000 erytrocytes). (a) A 7-year-old healthy female *M. mulatta.* Its mean reticulocyte count is 10.5/1000 RBC. (b) A 7-year-old healthy male *M. mulatta.* Its mean reticulocyte count is 10.9/1000 RBC. Each datum point on a graph is an independent count of a given day. The reticulocytes were supravitally stained with brilliant cresyl blue and enumerated by visualization under light microscopy at oil immersion magnification. (From Harne, O.G. et al., *J. Lab. Clin. Med.,* 30, 247, 1945.)

and a mean Rbc count of $5.10 \times 10^6/\mu L$ along with an absolute reticulocyte count of $44,400/\mu L$. Younger monkeys were similarly studied with a blood sampling schedule of every other week for 12 weeks (Table 1). Males less than 3 years old (24–32 months of age, n = 36) presented a reticulocyte count of 0.59% (s.d. ± 0.45) with a $5.70 \times 10^6/\mu L$ mean erythrocyte count and an absolute reticulocyte count of $33,600/\mu L$. Two hundred and twelve observations were conducted. The profile for females of the same age range (n = 22, 126 total testings) was a mean reticulocyte count 0.72% coupled with an average erythrocyte count of $5.57 \times 10^6/\mu L$. The absolute reticulocyte count was $40,100/\mu L$ of blood. A set of monkeys older than 3 years (36–60 months) were also followed for the same duration. (It is seen that this group has an age range that is similar to but slightly more restricted than the cohort that was followed for 40 weeks.) Males of this set (n = 38) maintained a mean reticulocyte count of 0.68% (s.d. ± 0.46) and a mean erythrocyte count of $5.85 \times 10^6/\mu L$ (219 observations). The absolute reticulocyte count was $39,800/\mu L$ of blood. In comparison, the female counterparts (n = 34) displayed a reticulocyte representation of 0.66% (s.d. ± 0.45) (201 samplings). The red cell count was $5.60 \times 10^6/\mu L$ while the absolute reticulocyte count totaled 37,000 reticulocytes/μL of blood. In overview of the analyses conducted in this investigation, it is seen that the male members of the first group of monkeys (n = 5) had a significantly higher erythrocyte count than the females (males 5.54 million/μL versus females 5.10 million/μL) while the proportional representation of the reticulocytes was equal in both sexes (♂0.91% versus ♀0.87%). The absolute reticulocyte counts were apparently dissimilar in the two sexes (males $50,400/\mu L$ versus females $44,000/\mu L$) due to the larger population of Rbc in males. The observations of the males and females more than 3 years old presented a comparable relationship. The percentages of reticulocytes that comprise the total population of erythrocytes in circulation are the same in both sexes even though the erythrocyte count is significantly higher in the males. The proportion of circulating reticulocytes to mature red cells is stable in healthy adults.

Reticulocyte counts (along with complete blood counts) were performed on 16 adult (11–22 years old) healthy, female nonpregnant *Macaca mulatta* before and approximately 15 minutes after intramuscular injection of ketamine hydrochloride (Ketaset) (Table 1, Bennett et al. 1992). Blood was obtained from the femoral vein. The subjects were restrained by two handlers during phlebotomy; the monkeys were conditioned to handling and appeared calm during the bleeding procedure. The analyses revealed statistically significantly decreased erythrocyte counts, hematocrits and hemoglobin concentration following the administration of ketamine when compared with the pre-ketamine sample (pre-ketamine: Rbc $5.97 \times 10^6/\mu L$, Hct 44.7%, Hb 14.1 g/dL; post-ketamine: $5.50 \times 10^6/\mu L$, 41.1%, 13.0 g/dL, respectively). The reductions were attributed to ketamine-induced reversal of the alarm response associated with physical restraint. The decreases in the major erythroid parameters (Rbc, Hct, and Hb) were considered to be caused by the redistribution of the erythrocytes to the spleen. An influx of fluid into the vascular space was also hypothesized. The mean reticulocyte count, however, was not significantly modified (pre-ketamine 1.5%, post-ketamine 1.7%). The absolute reticulocyte counts derived from the reported data were as follows: pre-ketamine 89,550 reticulocytes/μL, post-ketamine

93,500 reticulocytes/μL. Since the proportional representation of the reticulocytes in the population of circulating erythrocytes was not significantly altered by the experimentation, this stable relationship could be used as one indicator that suggests reticulocytes are not selectively treated one way or another differently from mature red cells when erythrocytes are sequestered in or released from the spleen.

The experience of Usacheva and Raeva (1963) with reticulocyte levels in the rhesus monkey differs from that of other workers. They analyzed the morphologic blood picture of 45 singly caged monkeys (36♂ and 9♀) that were 1–2 years old. The subjects were deemed healthy. This population of macaques presented a mean reticulocyte count of 7.8% for a mean erythrocyte count of 5.5 million/μL and mean 12.9 g Hb/dL. This elevated level of reticulocytes may have been due to the presence of an unrecognized compensated anemia. The report that the blood picture of these monkeys demonstrated polychromasia of the erythrocytes, aniso-cytosis due to the presence of microcytic, and macrocytic red cells along with the frequent occurrence of normoblasts strongly supports this premise. The occur-rence of a large number of polychromatophilic erythrocytes in blood that has a high reticulocyte count is indeed a consistent observation as well as an implication that the reticulocyte count was accurate.

Reticulocyte levels of *Macaca fascicularis* the cynomolgus *monkey* (captive-bred, 3–5 years old) were determined in the absence of as well as following intra-muscular injection of ketamine (Table 1, Kim et al. 2005b). The mean level for males (n = 19) was 1.73% for the untreated state (erythrocyte count 5.26×10^6/μL, absolute reticulocyte count 91,000/μL) and 2.02% when the sedative was adminis-tered (erythrocyte count 5.56×10^6/μL, absolute reticulocyte count 112,300/μL). For the females (n = 16), the reticulocyte count in the unmedicated condition was 1.60% (erythrocyte count 5.13×10^6/μL, absolute reticulocyte count 82,100/μL) while in the anesthetized state it was 1.69% (erythrocyte count 5.35×10^6/μL, absolute retic-ulocyte count 90,400/μL). The difference in percentage of levels of reticulocytes between drug-free and ketaminized assessments (obtained within 30 minutes of injection) was not statistically significant in either sex. The erythrocyte counts were significantly different for both sexes between their respective medication-free and medicated levels. The erythrocyte count associated with ketamine was seen to be increased. This reaction differs from the rhesus monkey in which ketamine admin-istration is seen to result in a reduction of the Rbc count (cf. section of the text that deals with ketamine).

The mean reticulocyte level for 333 male *Macaca fascicularis* between 2 and 5 years old maintained at the 67 member companies comprising the Japan Pharmaceutical Manufacturers Association was reported to be 0.44%, an abso-lute count of 25,920 reticulocytes/μL for their erythrocyte count of 6.48×10^6/μL (Matsuzawa et al. 1993). The reticulocyte levels for mature cynomolgus macaques (n = 27♂, 15♀) performed multiple times for each participant were in the same range in Schuurman and Smith's (2005) investigation. The representation for the males was 0.4% yielding an absolute 30,000 reticulocytes/μL while the mean for females was 0.55%, an absolute reticulocyte count of 33,000 reticulocytes/μL. A more contempo-rary study of reticulocytes in *Macaca fascicularis* by Wang et al. *(*2012) identified a mean 1% reticulocytes along with an absolute reticulocyte count of 62,890/µL in the blood of 35 males and 1.29% of these immature red cells coupled with an absolute reticulocyte count of 79,050/µL in 37 females. The monkeys were 2.5–3.5 years old and were cited as juveniles. The differences in both percentage levels and absolute counts between the sexes were statistically significant. This inquiry was noteworthy in that the electronically quantitated reticulocytes were identified as manifesting low, middle, or high fluorescence. This presumably indicates that the relative amount of ribonucleoprotein is greatest in the red cells emitting the high fluorescence because it binds with the maximum amount of the fluorescent label. This electronic fluorescence-activated method of identifying and estimating the amount of ribonucleoprotein in an erythrocyte demonstrated that approximately 6% of the total reticulocyte population emitted high fluorescence (the most immature cells), about 7%–10% presented middle fluorescence and the remaining ~85% fluoresced the least (the oldest "labeled" cells with the least ribonucleoprotein).

The incidence of reticulocytes in the circulation of normal monkeys has a relatively broad range, roughly 30,000–90,000/µL. Whether this represents an actual variation among animals, or is due to differences in the red cell levels of individuals because of physiologic variation of the volume of blood or the erythrocyte count, or is due to analytical technique is not identifiable.

The previously discussed five male 3–4-year-old cynomolgus monkeys of Kim et al. (2005a) that underwent stress (documented by elevated serum cortisol levels) as a result of transport from Japan to Korea and instillation into a new domicile demonstrated unpredicted changes in their reticulocyte counts. The erythrocyte and Hb levels were significantly elevated in the initial blood sample and thereafter receded to a lower level throughout the 35-day investigation. The blood was obtained from ketaminized subjects (10 mg/kg). The reticulocyte counts were performed by laser optical techniques and were interesting in that they were low at the initial sampling and second sampling 1 week later (day 0 absolute reticulocyte count 13,300/µL, Rbc 6.65×10^6/µL, reticulocyte percentage 0.2%; 1 week later, absolute reticulocyte count 23,280/µL, Rbc 5.82×10^6/µL, reticulocyte level 0.4%). Thereafter, the red cell count stabilized in the 5.66–5.37 million/µL span and the reticulocyte count increased to the "normal range." The absolute reticulocyte occurrence was 56,600–87,516/µL and represented 1.0%–1.56% of the erythrocyte count. Thus, in this study, a diminished absolute reticulocyte count was associated with a stressful event.

CHAPTER 24

Anemia

Anemia (which literally means without blood) can be defined as a condition whereby the total mass of erythrocytes present in a given individual is reduced to a level that is inadequate to sustain the normal, active life of the subject. Since the primary function of the erythrocyte is to transport and dispense oxygen, anemia results in a deficiency in the oxygenation of the body tissues. It is consequently appreciated why an anemic individual is less likely to conduct the activities of daily living with the same vigor as a healthy counterpart (e.g., to search for food, to compete for mates, or to care for infants). The oxygen-poor environment of the tissues of an anemic monkey can also be anticipated to reduce its capacity to combat infections and repair damaged tissues.

The occurrence of anemia is hematologically recognized by the demonstration of a decrease in one or more of the following parameters: the erythrocyte count, hematocrit, and hemoglobin concentration of the blood. Anemia has also been specifically defined as an erythrocyte count, hematocrit percentage, or hemoglobin concentration that is two standard deviations below the mean or the 2.5th percentile of the normal distribution of a healthy, iron-replete population (Adams et al. 2014). Hemoglobin concentration of blood is considered the most sensitive parameter to identify an anemia (Adams et al. 2014). It is quantitated by direct spectrophotometric determination of the amount of Hb present in a sample of blood typically after test tube conversion of the Hb to cyanmethemoglobin. Conversely, the hematocrit can be viewed as an indirect or derived measurement since in a modern laboratory it is mathematically derived by a formula that uses the prior determined erythrocyte count and mean red cell volume. The older, original technique of centrifugation of a sample of blood and visual measurement of the resultant column of red cells to establish the hematocrit has an inherent error of the inclusion of plasma trapped in between the cells yielding an imperfect quantitation of the volume of packed cells. This topic is considered at another site. The red cell count, though conceptually and historically the most straightforward indicator of anemia ("low erythrocyte count"), has the problem of not indicating the degree of loss of oxygen-carrying capacity because it does not indicate alteration or variation of cell size, or degree of hemoglobinization of the individual red cells. A subject with a "normal" red cell count can have a concentration of Hb or Hct level that indicates the presence of the anemic state. Adams et al. (2014) have noted that in their group study of the development of anemia in adult cynomolgus monkeys following phlebotomy of specific portions of a subject's total blood volume, the hemoglobin level was twice as likely to indicate the fall of the indicator to recognized anemic levels than was the Hct. The erythrocyte indices (which are mathematically derived from the three primary erythrocytic quanta, i.e., Rbc, Hct, Hb) assist in the recognition of the anemic status and/or promote the identification of the type of anemia that is manifested by the subject under scrutiny. Thus, some hemolytic anemias are associated with macrocytic erythrocytes (i.e., exhibiting an increased MCV), while monkeys with iron-deficiency anemia present hypochromic erythrocytes in their circulation (Figure 39, Wills and Stewart 1935). Hypochromic erythrocytes are given this description because when examined in a Wright-stained dry film smear, their area of central pallor is larger than that of normal erythrocytes. Such cells contain less than the normal quantity of hemoglobin whose presence in the erythrocyte is the reason for the cell's affinity for red stain eosin that is a component of Romanowsky stains. These cells have a decreased MCH.

The evaluation of the morphology of the erythrocytes in an anemia (as well as under normal conditions) is typically accomplished by a microscopic examination of a Wright (or other Romanowsky-type dye mixture)-stained dry film smear of blood. This appraisal aids in the documentation of the presence of the anemic state, the cytologic characterization of the anemia, and in some circumstances permits the identification of the cause of the condition (e.g., parasitization of the erythrocytes). Some anemias result in the generation of particular alterations of the morphology of the red blood cells and their recognition can offer an insight into the disease process; for example, an increased variation in the size of the erythroid cells (anisocytosis) could be due to a greater than normal incidence of immature cells in the circulation. Automated methods of performing erythrocyte counts and related parameters can also assist in the determination of the degree of anisocytosis with the calculation of the red cell distribution width, that is, RDW%, which is the coefficient of variation of the MCV. Supravital staining can be used to identify reticulocytes. When necessary, further morphologic assessments are obtained from the study of smears or histologic sections of bone marrow and sections of organs such as the spleen.

The documentable anemic state presents itself when a loss or destruction of normal or abnormal erythrocytes prevails to the extent that an individual's capacity for recuperative erythropoiesis is exceeded and the imbalance becomes evidenced by abnormal hemogramic values. Conversely, a verifiable anemia can also develop in circumstances where the hemopoietic marrow is incapable of sustaining a normal effective level of activity and a deficit in erythrocytes ensues. The concept of a compensated anemia can also be encountered. In this case, the subject is able to maintain a normal Rbc count, hematocrit, and hemoglobin concentration despite the presence of an erythrocytic disturbance because of the ability to generate a sustained, enhanced level of hemopoietic activity.

The etiologies for the development of an anemic state are numerous. The causation may be straightforward or subtly induced, and the deficit may reside within the erythrocyte itself or may be the result of disadvantageous conditions imposed upon a monkey's erythrocytes or its erythropoietic processes. The field of biochemistry is

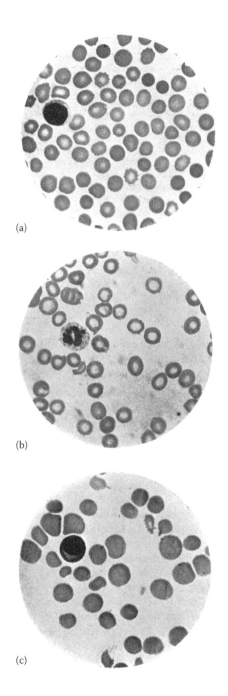

(a)

(b)

(c)

Figure 39   (*Continued*)

Figure 39 (Continued)   *Macaca mulatta.* Photomicrographs of stained dry film smears of blood from rhesus monkeys *Macaca mulatta.* They were obtained from one of the early studies of the hemopoiesis of this monkey as observed under different experimental conditions. (a) The blood of a normal, healthy subject. (b) The blood obtained from a monkey subjected to repeated phlebotomies yielding a microcytic hypochromic anemia. The increase in the central pallor (hypochromasia) of the erythrocytes is apparent. (c) The blood obtained from a subject that was placed on an experimental diet that was extremely low in protein, devoid of animal protein, and though supplemented with vitamins A, C, and D, was low in vitamin B complex vitamins. Iron was added to the diet. This treatment yielded a macrocytic anemia within periods of 3–9 months (e.g., an erythrocytic mean corpuscular diameter that increased from 6.7 to 8.1 µm). The stained erythrocytes appeared full of hemoglobin and did not exhibit a central pallor. Megaloblasts were always present in the blood at the peak of the anemia. Megaloblasts and a marked hyperplasia characterized the bone marrow. The presented blood film illustrates the reduction of the number of circulating erythrocytes, their increased size, as well as the absence of the central pallor (the latter feature is now recognized as due to the increased thickness or depth of the smeared cells). The nucleated cell in the field is a developing erythroid cell. The nucleus is intensely stained thereby preventing its identification as a normoblast or megaloblast. Original magnification ×570. (From Wills, L. and Stewart, A., *Brit. J. Exp. Path.,* 16, 444, 1935.)

a source of innumerable analyses that are applicable to the investigation of anemia (e.g., lead analysis, serum/tissue iron analyses, enzyme assays).

The detailed exposition of the anemias of the rhesus and cynomolgus monkeys is too vast a subject to be adequately treated by this text. That endeavor is well embraced by numerous texts (such as Schalm's Veterinary Hematology, 2010) and published scientific reports that are available from the usual academic sources. The purpose of this section of the text is to offer an encapsulated, abbreviated overview of the principal mechanisms that lead to the development of anemia in monkeys.

1. *Regenerative anemia*: Regenerative anemias are those in which the erythropoietic activity of the bone marrow is enhanced (i.e., normoblastic hyperplasia is present), and this response is quantitatively and qualitatively appropriate for the amelioration of the anemic insult. Such anemias are characterized by blood films in which the erythrocytes display polychromasia, reticulocytosis, macrocytosis, and anisocytosis. The recruitment of normoblasts into circulation may also occur. These responses are interpreted as indicators of enhanced erythropoiesis.

2. *Hemolytic anemia*: A hemolytic anemia by definition is one in which the life span of the erythrocyte is reduced. The abbreviation may be the result of a defect within the structure of the erythrocytes or because normal erythrocytes are exposed to an aberrant, disadvantageous milieu. Hemolytic anemias are often regenerative in nature as indicated by increased polychromasia, anisocytosis, and macrocytosis. Most intraerythrocytic parasites have the potential to cause an anemia and one of the parasites frequently associated with hemolytic anemia is the historically well-recognized malarial organism *Plasmodium* (Wills and Stewart 1935).

3. *Hemorrhagic anemia*: Hemorrhagic anemia by its title indicates that this form of anemia is due to loss of blood. A typical cause is gastrointestinal parasitism.

When the hemorrhagic incident(s) occur in an otherwise healthy individual, it is readily understood that the erythropoietic response permits this condition to be also included with regenerative anemia.

4. *Nonregenerative anemia*: Nonregenerative anemias (also termed *hypoproliferative anemias*) are those associated with a lack of an appropriate erythropoietic response in the bone marrow. This type of anemia is likely to have erythrocytes in circulation that are normocytic (normal MCV) and normochromic. A typical nonregenerative anemia is likely to have a low reticulocyte count, mirroring a lowered rate of generation erythrocytes by the bone marrow. Causes of nonregenerative anemia include chronic inflammatory conditions and chronic infectious diseases.

It is also feasible to include the anemia associated with leukemia in this category. Every type of malignant hemoproliferative condition, at least in its final stages, can be expected to develop anemia with ample concomitant evidence that the bone marrow is incapable of providing the ill subject with a normal erythroid picture.

When the bone marrow demonstrates erythroid hyperplasia (a greater than normal representation of normoblasts) and the development of erythroid cells is impaired resulting in an ineffective production of red cells and persistence of the anemic state, the condition would also be classified as nonregenerative anemia.

Wills and Stewart (1935) were among the very earliest to attempt to document some causes of anemia in the rhesus monkey and to classify the anemias on the basis of erythrocytic morphology. They demonstrated that chronic phlebotomy (two subjects) results in a diminished erythrocyte count and hemoglobin concentration of the blood, a reduction in mean cellular diameter, and an enhanced reticulocyte count. An increased volume of red bone marrow and normoblastic hyperplasia were found at autopsy. Experimental long-term infection with the malarial organism *Plasmodium knowlesi* similarly resulted in severe microcytic anemia with moderate anisocytosis, marked polychromasia, reticulocytosis, and circulating normoblasts. The monkeys were kept alive by giving them medication and relapses were associated with large numbers of the organism in the circulation. At *postmortem*, large amounts of malarial pigment were observed in the red pulp of the spleen. In their other studies, deficient diets typical of the poor residents of Bombay (Mumbai), India generated anemias that were macrocytic as well as megaloblastic.

BLOOD LOSS

A straightforward phlebotomy-based program to produce an anemia in the rhesus monkey was designed by Wolcott et al. (1973). They bled 2–5-year-old males (n = 20) 10 mL/kg of body weight three times per week for a period of 30 days (a weekly loss of ~50% of the whole body volume of blood). The reduction in Rbc, Hct, and Hb was definite. Initial pre-phlebotomy levels were Rbc 5.2 × 10^6/μL, Hct 43%, Hb 13.8 g/dL, while the final values at 30 days were Rbc 2.9 × 10^6/μL, Hct 23%, Hb 6.1 g/dL (Figure 40). An increase in the reticulocyte count was generated by this schedule, as would be expected. Interestingly, it peaked at around day #17 and decreased thereafter until the termination of the experiment when it reached

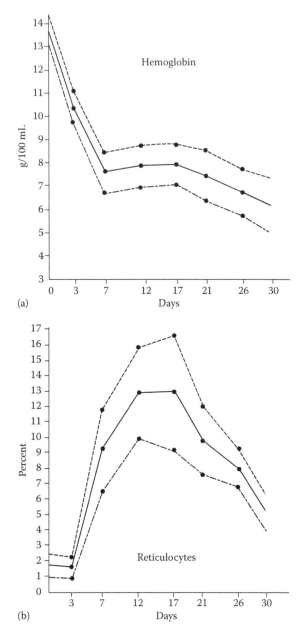

(a)

(b)

Figure 40 *Macaca mulatta*. Hematologic results of phlebotomies of rhesus monkeys, 10 mL/kg body weight, three times per week during a period of 30 days. The subjects were 20 males, 2–5 years old. At the end of the 30 days, the hemoglobin had fallen from a mean 13.8 to 6.1 g/dL. The hematocrit receded from a mean of 43% to a mean of 23%. The reticulocyte count increased from 1.6%, peaked at 16.5% (day 17), and then receded to 4.6% by the end of the study. The broken lines indicate 1 SD. (From Wolcott, G.J. et al., *Lab. Anim.*, 7, 297, 1973. With permission from SAGE.) *(Continued)*

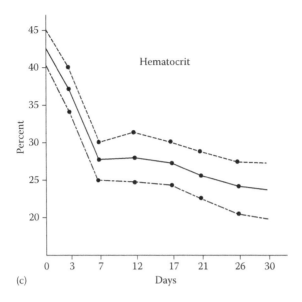

(c)

Figure 40 (Continued) *Macaca mulatta.* Hematologic results of phlebotomies of rhesus monkeys, 10 mL/kg body weight, three times per week during a period of 30 days. The subjects were 20 males, 2–5 years old. At the end of the 30 days, the hemoglobin had fallen from a mean 13.8 to 6.1 g/dL. The hematocrit receded from a mean of 43% to a mean of 23%. The reticulocyte count increased from 1.6%, peaked at 16.5% (day 17), and then receded to 4.6% by the end of the study. The broken lines indicate 1 SD. (From Wolcott, G.J. et al., *Lab. Anim.,* 7, 297, 1973. With permission from SAGE.)

4.6% (Figure 40). This configuration of the reticulocyte curve can be interpreted as an initial robust, energized erythropoietic response to the anemic stimulus while the subsequent diminution of the reticulocyte count was due to a decreased capability to sustain maximal erythropoiesis in light of the continued loss of red cells, plasma protein, and serum iron. The absolute reticulocyte at the onset of the study totaled 83,000/μL while it was 133,000/μL at the end, an increase that yielded less than twice the basal level of reticulocytes.

An earlier study of the effects of blood loss upon the erythrocyte count and hemoglobin level in the rhesus macaque was undertaken by Krise and Wald (1959). Their subjects (n = 5♂ per test group, ~3 years old, 1.8–3.6 kg) that were phlebotomized 60 mL during one interval (~37% of the circulating blood volume, on the assumption of a blood volume–body weight ratio of 61 mL/kg) demonstrated a resultant drop in Rbc and Hb, which returned back to normal levels by 28 days post-phlebotomy. Animals that were subjected to either 4 or 10 mL venisections (2.5% or 6% of the estimated blood volume) weekly for a period of 8 weeks did not significantly alter their hemic profile during the duration of the experiment. The cohorts that were bled 30 or 40 mL (18% and 24% of the estimated blood volume) each week for 8 weeks did not return to their initial levels during the 8 weeks of bleeding and analysis.

Among the definite observations made here are that a normal monkey can sustain an acute withdrawal of approximately 30% of its blood volume and the bone marrow can compensate for the loss within 1 month of the event. A weekly loss of either 4 or 10 mL of blood for 8 weeks will not significantly modify the blood picture (although the 10 mL losses do result in a blood picture that is typically ≥90% of the basal level).

A yet another more recent investigation of the effects of repeated phlebotomy of *Macaca mulatta* was conducted by Mandell and George (1991). In this assessment, healthy mature nonpregnant females 6.5–10 years old (mean 8 yr) weighing 4.8–10.0 kg (mean 6.3 kg) were phlebotomized either 5% or 10% of their calculated blood volume (61 mL/kg) per week for 10 weeks or until their Hb level diminished to ≤10 g/dL. Each test group was comprised of six animals and six other subjects served as controls. The majority (8/12) of the bled monkeys became anemic (≤10 g Hb/dL) after 30%–70% (mean 49%) of their blood was removed (at 5% or 10% per week). Microcytosis, hypochromasia, and increased erythrocyte zinc protoporphyrin concentration all recognized indicators of iron deficiency developed later. The mean hemoglobin concentrations in both groups differed significantly from the controls at 3 weeks, persisted to decline, and reached a plateau (both cohorts) during the interval of 8–10 weeks. Of potential note was the observation that of the 12 experimental animals eight had reached the anemic level (≤10 g Hb/dL) by week 10. Five of the latter eight monkeys had decreased to that level by week 5. The rate of anemic response to blood loss as measured by the decrease in the level of hemoglobin in the blood, at least as seen in this group of adult female monkeys, displays considerable variation. The subjects that were more or less susceptible to develop an anemia were not predictable. This attribute, should it prove to be a genuine characteristic of laboratory rhesus monkeys, could be of significance in some conditions or investigations. Additional findings and conclusions that were generated by this experiment would be the following. As would be predicted, the group that was phlebotomized 10% of its blood volumes per week demonstrated a more rapid mean decrease in Hb than the 5% group. Interestingly, both cohorts' mean hemoglobin leveled off at roughly equal values at the 10th week. It was also seen that the 10% phlebotomy was associated with a significant decrease in the MCV and MCH by week 5. A lesser decrement of these indices was obtained by the 5% group, which failed to attain significance throughout the test period. The observed results in this investigation are believed to be due primarily to the loss of iron but other factors such as loss of protein can be expected to have played a role. Mandell and George (1991) point out that in chronic blood loss attributable to repeated phlebotomy, the decrease in the concentration of hemoglobin in the blood is the first major erythroid parameter to present a significant change. Decrements in serum iron concentration and saturation of transferrin were the next earliest indicators of iron deficiency. These were followed by reductions in MCV and MCH as indicators of developing iron deficiency. And as cited earlier, although not identifiable by the level of hemoglobin or other erythrocytic parameters some healthy monkeys, at least in this study, apparently have sufficient iron stores to sustain a greater loss of blood than others before demonstrating indicators of iron deficiency.

The standard hematologic profile that is induced by an iron-deficient diet combined with repeated bleeding is described as a microcytic, hypochromic anemia with increased anisocytosis and poikilocytosis (Huser 1970; Wixson and Griffith 1986). In the pathogenesis of iron-deficiency anemia, hemoglobin synthesis is more deficient than erythrocyte production (Wixson and Griffith 1986). This conclusion has also been reached by Mandell and George (1991). Wixson and Griffith further state that in such treated animals a maturation arrest takes place in the late rubricyte stage (polychromatophilic normoblast) so that while these cells are awaiting hemoglobin synthesis further cell division takes place. This yields daughter cells that are smaller as well as deficient in hemoglobin. The bone marrow is characterized by a predominance of late rubricytes and metarubricytes (late polychromatophilic normoblasts).

Intestinal parasites that cause bleeding (and hence a potential cause of anemia due to loss of blood) were identified in feral *Macaca fascicularis* on Sumatra Island, Indonesia (Takenaka 1981).

A study was designed to evaluate the amount of blood that can be obtained from healthy adult male and female *Macaca fascicularis* for four consecutive weeks with minimal effect on their well-being (Adams et al. 2014). The monkeys were phlebotomized under ketamine anesthesia of 7.5%, 10%, 12.5%, 15%, or 17.5% of their total blood volume once weekly for 4 weeks and then subsequently hematologically assessed each week for the following 4 weeks. Four males and four females were assigned to each collection group. The overall results of the investigation indicated that both males and females tolerated removal as much as 15% of the total blood volume per week for the test duration of 4 weeks. (This latter level of blood loss resulted in mild anemia at one or more time points in three of the eight animals along with single episodes of emesis in two of them in the 24-hour period following a venisection.) Removal of 17.5% of the blood volume weekly for 4 weeks yielded a reduced level of the major erythroid parameters that did not return to original levels until 2–3 weeks after the series of bleedings (four venisections). This group exhibited a relatively high incidence of emesis or anorexia within 24 hours of a phlebotomy. Four of eight subjects presented two or more of such episodes. The mean of the principal erythroid parameters (Rbc, Hct, Hb, males, and females calculated independently) of the groups that were phlebotomized the least (7.5% of the blood volume per phlebotomy) were usually lower than the original value until the second week after the last phlebotomy. The investigators of this study have concluded that the results of their work are in accord with the opinion of other workers who hold that 1–3 weeks of rest may be required for the hematologic parameters to return to normal after 7.5%–20% of the blood volume is removed from an individual during a 24-hour period. All levels of blood withdrawal were associated with reticulocytosis and macrocytosis until 2 weeks after cessation of the phlebotomies.

Certain biochemical values have a specific relationship with anemia that is induced by significant blood loss. The most familiar parameter is the concentration of iron in the serum of the blood (i.e., serum iron). The iron in serum is bound to transferrin, the only protein in the plasma that transports iron. It is a glycoprotein and apparently has (species-dependent?) one or two iron-binding sites. Hartwig et al. (1958) have reported that a base of 455 determinations of rhesus monkeys have generated a mean plasma iron of 185.3 µg/dL, s.d. 46 µg. All samples were obtained at the same time of day. Huser (1970) cites 101 µg/dL (range 64–124, n = 9) as the normal average level of serum iron in the rhesus monkey. Sood et al. (1965) have reported a mean 172 µg/dL (range 83–308) for their 12 young growing monkeys (≥9♂, weight between 2 and 4 kg). The mean serum iron of mature nonpregnant female *Macaca mulatta* has been reported to be 171 µg/dL (n = 9) (Spicer and Oxnard 1967). The concentration of serum iron in man is ~40–160 µg/dL. The previously discussed bled male rhesus monkeys (10 mL/kg body weight, 3 times/week, for 4 weeks) of Wolcott et al. (1973) had an initial mean serum iron of 189 µg/dL that fell to 57 µg/dL by the end of the investigation. In another also previously noted circumstance, repeatedly phlebotomized mature nongravid female rhesus monkeys (5% and 10% of the blood volume, once per week × 10 weeks) developed a significant reduction of their serum iron during the treatment (Mandell and George 1991). Their mean pre-venisection level of serum iron was in the range of 175 µg/dL and by 7 weeks of bleeding it had receded to ~50 µg/dL. In contrast to phlebotomy-associated depletion of serum iron, an incidental example of elevated serum iron has been cited in an infant rhesus monkey that developed an autoimmune hemolytic anemia due to experimental injection of four neurophysiologic agents into its brain. The level of serum iron rose to ~400 µg/dL (Suzuki et al. 2000); cf. section "Experimentally Induced Autoimmune Hemolytic Anemia."

Concurrent with the reduction of serum iron in phlebotomy-induced anemia of *Macaca mulatta*, an understandable decrease in the level of saturation of transferrin with iron is also identifiable. The level of saturation of transferrin with iron was 48% in Wolcott et al.'s (1973) normal monkeys (i.e., prior to venisections). Due to the subsequent paucity of iron following the serial bleedings, it diminished to 11%. An equivalent fall was similarly obtained in Mandell and George's (1991) phlebotomized macaques (following 7 weeks of bleeding). The saturation range for normal humans is 20%–50% (Kjeldsberg et al. 1989). A third evaluation is the total iron-binding capacity. It is the concentration of iron that is necessary to saturate the iron-binding sites of transferrin in a sample of serum. It is a measure of the concentration of transferrin in the serum. It would be anticipated that in an iron-deficiency anemia (as induced by repeated phlebotomy) that the proportion of transferrin in the serum that is carrying iron would be decreased from normal. Thus when the iron-binding capacity of serum from a subject with that condition is tested, a greater than "normal" amount of iron would be expected to bind to the transferrin in the sample. In addition, it can also be postulated that the amount of transferrin may be increased as a compensatory mechanism initiated by the body to enhance its binding to all available iron. Wolcott et al.'s monkeys had a mean total iron-binding capacity of 373 µg/dL prior to the series of bleedings. After 4 weeks of phlebotomies, the iron-binding capacity had risen to 505 µg/dL. The normal iron-binding capacity for *Macaca mulatta* as reported by Huser (1970) is a mean 411 µg/dL (range 330–514, n = 9). Sood et al. (1965) obtained a level of 441 µg/dL (range 275–547) for 12 healthy young monkeys. For man, the normal value is 250–410 µg/dL (Kjeldsberg et al. 1989).

Three 1.5-year-old rhesus monkeys (1♂, 2♀) that were maintained on a purified diet that was devoid of iron for 17 months were compared with two male control monkeys of the same age that were given the same diet, to which iron was added,

revealed the development of classical iron-deficiency anemia in the subjects administered the experimental diet (Munro 1987). Worthy of note was the observation
that the erythrocyte counts of both groups were essentially the same at the end of
the experimental period, that is, iron-deprived monkeys 5.9×10^6 Rbc/µL versus
6.0×10^6/µL for the controls. On the other hand, the hematocrits for the respective
two groups were 27%–29% and 46%–49%, while the hemoglobin level for the iron-
deficient monkeys was 7 g/dL and that for the normal animals was 14 g/dL. The
MCV was impressively reduced in the deficient subjects, 37 fL as opposed to 67 fL
for those maintained on the normal diet (Table 1). The other iron-related biochemical values were consistent with what would be expected to be observed in the two
test groups. The serum iron, serum iron-binding capacity, and transferrin saturation,
respectively, for the anemic monkeys were 12 µg/dL, 547 µg/dL, and 3%. The serum
iron of the control monkeys was 192 µg/dL, the serum iron-binding capacity was
492 µg/dL, and the transferrin saturation was 38%.

In a population of mixed-age (2 months old to ~10 years old) colony-bred *Macaca
fascicularis* (n = 35♂, 31♀), the mean serum iron was 99 µg/dL (range 30–154) for
males and 110 µg/dL (range 34–184) for females (Giulietti et al. 1991). The saturation of transferrin with iron for these long-tailed macaques in the males was 31%
(range 9%–61%), while in females it was 34% (range 11%–56%). The mean total
iron-binding capacity was 317 µg/dL (range 258–387) in males and (range 238–461)
in females.

The serum iron levels of 15 laboratory-bred and 15 captured ~2-year-old male
Macaca fascicularis were 97 and 110 µg/dL, respectively. The difference was not
significant. The average values for females of the same origins and number of test
subjects were 231 µg/dL for the laboratory-bred monkeys and 187 µg/dL for the
captured long-tailed macaques (Bonfanti et al. 2009). The difference in this instance
was significant, $P \leq 0.05$.

ANEMIA DUE TO DEPRIVATION OF DIETARY PROTEIN

It would be anticipated that if a rhesus monkey were deprived of protein in its
diet, the monkey would become anemic (along with other conditions). This premise was experimentally analyzed by Sood et al. (1965). Twenty-four young growing *Macaca mulatta* (all males except three subjects, weight between 2 and 4 kg)
were divided into two groups. Seventeen animals were rendered protein deficient
by feeding them a protein-deficient diet and seven monkeys served as controls. The
subjects were tube-fed 100 cal/kg of body weight per day. The diets were identical
(isocaloric) for both sets except the diet for the control monkeys was rich in protein
because it contained 15% casein. All animals were given 10 mg of ferrous sulfate as
well as some other vitamin and mineral supplements. This presumably insured that
any results would not be related to a deficiency of iron. After 8–10 weeks on the protein-deficient diet, all monkeys except one developed a moderately severe anemia.
Each major erythroid parameter (Rbc, Hct, Hb) was significantly lower. The hemoglobin concentration, quantitated by the cyanmethemoglobin technique, declined

from a mean 12.6 to 10.4 g/dL, P < 0.001. Four animals were maintained on the deficient diet for an additional 5 weeks and presented a continued fall in the red cell count, hematocrit, and hemoglobin concentration. Two of the anemic animals were placed on the protein-rich (control) diet for 8 weeks, and their hemogramic values returned to normal. The anemia encountered in the deficient monkeys was normocytic and normochromic. The MCV and MCH were slightly elevated but did not attain statistically significant levels. Bone marrow samples were aspirated from the iliac crest. The Leishman-stained smears revealed normoblastic maturation (and no abnormalities in the other hemopoietic cell lines). The M:E ratio was slightly increased. This was suggested to be an indication of reduced erythropoiesis. The serum iron displayed a progressive fall in the protein-deficient diet-treated monkeys. It declined from a basal 172 to 111 µg/dL, P < 0.01 following 8 weeks of treatment. This change could be reversed by placing the monkeys on the protein-rich control diet. The serum total iron-binding capacity similarly demonstrated a definite progressive reduction in the treated subjects, initial mean 441 µg/dL and final at 8–10 weeks a mean 285 µg/dL, P < 0.001. The absorption of radioactive tracer iron out of the gastrointestinal tract was also deficient in the anemic cohort. The average value for absorption in the deficient monkeys was 40% as compared to their basal, initial value of 51%, P < 0.02. An unsurprising observation was that the monkeys maintained on the protein-deficient diet had diminished levels of total serum proteins and albumin fraction.

FERAL ANEMIC COHORT IN SUMATRA

A cohort of wild *Macaca fascicularis* crab-eating monkeys residing in Gunung Meru near Padang in West Sumatra, Republic of Indonesia was documented to be probably displaying a hemolytic anemia perhaps compounded with other causes for anemia (Table 1, Takenaka 1981). The group exhibited mean low major erythroid values: Rbc 4.41 and 4.30 × 10^6/µL, Hct 34.7% and 33.2%, and Hb 9.6 and 9.0 g/dL for males and females, respectively (n = 17♂ and n = 23♀). The erythrocyte indices were uniformly lower in both sexes in comparison with the feral colonies of monkeys of Bali Island, Republic of Indonesia (e.g., the Sangeh group, Table 1) that were also subjects in the investigation. The anemic monkeys had an elevated plasma level of transferrin (this carrier of iron is elevated in iron deficiency). In addition, some of the anemic monkeys had no serum haptoglobin (when hemolysis occurs it binds to the liberated hemoglobin and the complex is degraded in the liver). The absence of haptoglobin is thus consistent with the occurrence of hemolysis.

MALARIA AND BABESIOSIS

It is well known that intraerythrocellular parasites have an adverse impact upon the invaded erythrocytes and also cause their destruction. They are thereby a cause of hemolytic anemia. Malaria is the best known, universally recognized parasite of monkeys' (and man's) red cells. The erythrocytes of some of the rhesus monkeys occurring in Viet Nam were recognized to be parasitized by *Plasmodium* (the malarial organism) as early as 1911 (Mathis and Leger). The latter workers published illustrations of Giemsa-stained cells displaying the various developmental stages of the organism (Figure 41). They reported that 5 of 40 studied subjects demonstrated the organisms and that none of the monkeys presented signs of illness. It is characteristic of a malarial infection to persist for extended periods, sometimes the life of the host as a chronic asymptomatic condition.

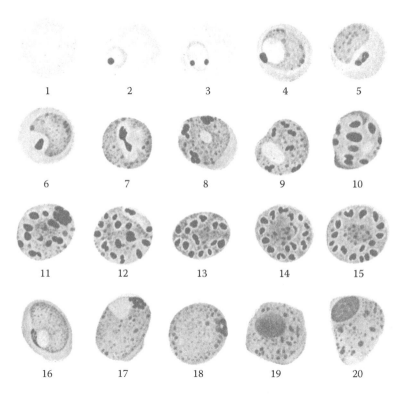

Figure 41 *Macaca mulatta*. Erythrocytes of *Macaca mulatta* parasitized by *Plasmodium* (malaria). (1) Normal erythrocyte. (2–3) Ring-form trophozoites. (4–6) Developing ameboid trophozoites. (7–15) Trophozoites evolving into a meront. At this stage of development, the amount of cytoplasm of the organism increases and its nucleus undergoes multiple asexual divisions. Once the process of cytoplasmic hypertrophy and nuclear division initiates, the organism is termed a meront or a schizont. The individual uninuclear parasites that are generated from the aggregated mass are called *merozoites*. They rupture the erythrocyte in which they develop and can then invade other red cells. (16–18) Macrogametocytes (female sexual forms). (19–20) Microgametocytes (male sexual forms). In many cases, the parasite is large enough to occupy the entire erythrocyte and it is difficult to identify the red cell. Giemsa stain. 1800×. (From Mathis, C. and Leger, M., *Annales de L'Institut Pasteur*, 25, 593, 1911.)

As Ameri et al. (2010) have recently reviewed, species of *Plasmodium* undergo a sexual development phase in the vector mosquito and asexual cycles in erythrocytes of the vertebrate host. Sporozoites delivered from an infected mosquito during feeding on the nonhuman primate are carried to the monkey's liver. Here, asexual developmental cycles termed *merogony* occur within hepatocytes. The resultant merozoites then invade the erythrocytes in which asexual cycles continue to occur (and also alternatively into sexual microgametocytes and macrogametocytes which are infective for the mosquito). The latter consequently permit infection of mosquitoes and transfer of the disease to other monkeys.

Small spherical or oval merozoites enter the circulation and invade the erythrocytes. The intraerythrocytic merozoites transform first into ring forms (ring trophozoites). This morphologic presentation is due to an eccentric placement of the nucleus and the presence of an unstained (white) vacuole in the center of the spherical or somewhat spherical trophozoite which prior to Romanowsky dry film staining of blood contained lipid material. The staining procedure extracts the lipid yielding the unstained region. The ring forms continue their development into larger, non-ring "ameboid" form trophozoites and finally into either asexual meronts or the sexual forms (male microgametocytes and female macrogametocytes). The trophozoites that evolve into intraerythrocytic meronts grow by increasing their amount of cytoplasm and the size of their nucleus. The nucleus undergoes multiple asexual divisions and it becomes visibly apparent how each newly arisen nucleus along with its surrounding cytoplasm may, upon segmentation of the cytoplasmic mass, give rise to a new parasite. Once the process of hypertrophy of the cytoplasmic mass and nuclear division initiates, the organism is considered a meront. The mass may alternatively be titled a schizont while the division of the mass into an aggregate of parasites is termed *schizogony*. (*Schizo* is the Greek stem for splitting or cutting; its application to a phenomenon of segmentation to yield a number of units is therefore mnemonically helpful.) The individual uninuclear parasites that are generated from this aggregated mass are called merozoites. The number of merozoites that develop in a given erythrocytic meront is one of the major differentiative characteristics that are employed to identify the species of *Plasmodium*. When the merozoites rupture the red cell in which they arise and enter the circulation, they are able to invade other erythrocytes. The development of intraerythrocytic *Plasmodium* was well illustrated by Mathis and Leger (1911, Figure 41).

The malarial parasites produce their greatest pathologic effects while schizogony is occurring in the blood. During the intraerythrocytic phase, the parasite digests the host hemoglobin. The intraparasitic free heme is converted to microscopically visible, insoluble β-hematin, which is called hemozoin or malarial pigment. In addition, the erythrocytes sensitized by antibodies produced during the infection may adhere to each other as well as to capillary endothelium thus impeding blood flow in the microcirculation. Such sludging of erythrocytes has been cited in macaques (Loeb et al. 1978). Malaria can be transmitted not only by mosquitoes but also congenitally as well as by the inoculation of parasitized blood.

Malaria was found to be "naturally occurring" in rhesus monkeys in East Pakistan (Schmidt et al. 1961). The majority of the cases in the latter study were thought to be due to *Plasmodium inui*. In two instances, a combined infection of the latter organism along with *Plasmodium cynomolgi* was tentatively identified. Wild, long-tailed macaques *Macaca fascicularis* in this part of the world have been described as commonly infected with *Plasmodium knowlesi*, *Plasmodium cynomolgi*, and *Plasmodium inui*. The latter organism is said to be the most common malarial species in cynomolgus monkeys (Ameri et al. 2010). In fact, the latter investigators made an unanticipated diagnosis of this parasite in an imported *Macaca fascicularis* from China. Some of the anemic *Macaca fascicularis* studied in Indonesia by Takenaka (1981) (discussed earlier) had a potential definitive cause for hemolysis as *Plasmodium* species was identified in their blood films. Current-day scientific literature offers many reports concerning malarial infection of the macaques' erythrocytes.

The tick-transmitted intracellular parasite *Babesia* invades the erythrocytes of macaques just as it similarly invades the red cells of man. It has been recorded as a naturally occurring infection in a colony of rhesus monkeys (Gleason and Wolf 1974; Loeb et al. 1978). The organism is a hematogenous sporozoan that is dissimilar to malaria in that it produces no pigment in the invaded erythrocytes and does not develop schizonts or gametocytes. Reproduction in the vertebrate host is asexual and occurs by binary fission. Pyriform, round, ameboid, rod-shaped, and ring stages are seen within the erythrocytes. They are thought to largely reflect the developmental process. In a great number of *cases*, perhaps the majority, the infection is latent, the monkey's red cells are devoid of parasites, and the blood picture is normal. On the other hand, severe babesiosis is associated with splenectomy as well as with anti-lymphocyte antibody therapy. In one investigation, *Macaca mulatta* experimentally infected with *Babesia microti* on days 1 and 110 followed by splenectomy on day 297 yielded a massive response of parasitized erythrocytes in the circulation that peaked 16–25 days after splenectomy and reached levels of 1.9×10^5–2.7×10^6/μL of blood. A mild to severe anemia was engendered along with a reticulocytosis and in some cases the presence of nucleated erythrocytes in the blood (Ruebush et al. 1981).

In another occurrence, two feral *Macaca fascicularis* were imported from Indonesia and housed in the Washington Regional Primate Center breeding colony for 7 and 12 years, respectively, and were asymptomatic. They had negative quarterly blood film analyses until they developed parasitemia after sustaining significant medical stress. In one case, the monkey contracted type D retrovirus infection, while in the other the long-tailed monkey sustained severe trauma leading to a necrotic infection. In both cases, *Entopolypoides macaci* (Babisiidae) was diagnosed on the basis of morphology from Wright-stained blood smears (Emerson et al. 1990, Figure 42). The rate of infection of the erythrocytes varied. The subject with the retrovirus infection had an invasion of 28% of its red cells at one point in the course of its combined diseases. In yet another situation, a 7-year-old rhesus monkey (originally imported from China and then housed at the Tulane National Primate Research Center) was intravaginally inoculated and became infected with simian human immunodeficiency virus (SHIV-RT) (Liu et al. 2014). It developed an anemia, fever, and anorexia in 2 weeks. The monkey attained the lowest erythrocyte counts and maximum lymphopenia 15 months after inoculation. At that time, Wright-stained

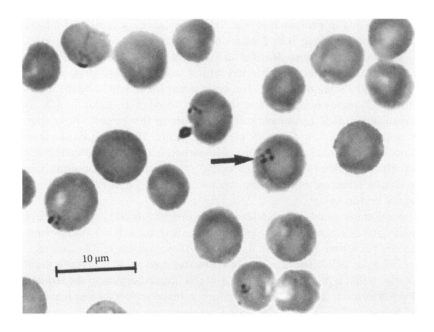

Figure 42	*Macaca fascicularis.* Babesia (*Entopolypoides macaci*) parasitzation of erythro-
cytes of a *Macaca fascicularis* monkey. The arrow indicates an infected erythrocyte. Other parasitized cells are also present in the field. Wright-stained blood
smear. (From Emerson, C.L. et al., *Lab. Anim. Sci.*, 40, 169, 1933.)

blood films were made and parasitized erythrocytes containing *Babesia microti*–
like organisms were identified. The parasitemia was persistent and varied from an
infection of >50% of the red cells at the peak of the lymphopenia to <5% of the red
cells at the end of the study 38 months after SHIV inoculation. The most common
organisms were piroplasms 1–4 μm in diameter in a signet ring, or pyriform structure that usually had one or two dark nuclei. The well-known classic tetrad forms of
merozoites (Maltese cross) were identified but were rare.

An interesting relationship has been observed between the frequency of circulating erythrocytes parasitized with *Babesia* and *Plasmodium cynomolgi* when both
organisms infect the same monkey. An individual *Macaca mulatta* imported from
China apparently had a latent but unrecognized infection with *Babesia microti*–like
(Wel et al. 2008). When this subject was experimentally intravenously inoculated
(along with three other monkeys) with *Plasmodium cynomolgi*, the latter three subjects presented a peak plasmodial parasitemia on day 12 after inoculation. However,
the single latently *Babesia*-infected monkey displayed an unexpected significantly
lower plasmodial erythrocytic infestation but also concurrently another hematozoon,
that is, *Babesia microti* like. The latter organism included the characteristic Maltese
cross-configuration (four pyriform-shaped trophozoites in a given Rbc). After the
(minimal) peak parasitemia on day 12, both parasitemias declined to concurrently
low levels. Of investigative interest was the fact that this subject was treated with chlo-
roquine on day 23 to cure its malaria. Two years later, this same macaque was once

again experimentally reinfected with *Plasmodium cynomolgi* and identical results recurred. Both the *Babesia* and *Plasmodium* infections were identified, and the plasmodial infestation was markedly suppressed in comparison with three other infected subjects (peak malarial parasitemia for the test subject 0.16% versus a mean 2.6% for the three control monkeys). Thus, in this study, the infection with the intraerythrocellular parasite *Babesia* was associated with a depressed intraerythrocytic invasion by *Plasmodium* when the monkey was inoculated with the latter organism.

HEMOLYTIC UREMIC SYNDROME

Hemolytic uremic syndrome (HUS) is an uncommon condition in man for which a potential experimental animal model has been developed in *Macaca mulatta* (Hillyer et al. 1995). The disease process is characterized by microangiopathic hemolytic anemia (an anemia caused by traumatic disruption of erythrocytes induced by a damaged endothelial lining of capillaries typically the renal glomeruli), platelet thrombi, thrombocytopenia, and renal dysfunction. It occurs primarily in children usually following an acute diarrheal episode acquired as a foodborne illness or from contaminated water and caused by a species of *Escherichia coli*. The syndrome has also been seen in adults following pregnancy, post-chemotherapy, and bone marrow transplantation (Loomis et al. 1989; Rabinowe et al. 1991). Monkeys developed this condition as a result of administration of high dose of cyclophosphamide (≥150 mg/kg) as opposed to those who received lower dosages and did not develop HUS (Hillyer et al. 1995). The classical clinical triad of an increased BUN (blood urea nitrogen)/creatinine, microangiopathic hemolysis, and thrombocytopenia developed within 1 week of administration of the drug (Table 14). The peripheral blood smears demonstrated morphologic findings consistent with a microangiopathic hemolytic anemia, that is, schistocytes, reticulocytosis, and nucleated erythrocytes.

Table 14   Clinical Laboratory Parameters of *Macaca mulatta* with Chemotherapy-Induced Hemolytic Uremic Syndrome (HUS)

	Hct	Platelets/μL	BUN	Creatinine
HUS *Macaca mulatta* Cyclophosphamide ≥150 mg/kg Values: days 5–8	31%	39×10^3	168 mg/dL	8.2 mg/dL
Control non-HUS *Macaca mulatta* Cyclophosphamide 100 mg/kg Values: days 5–8	35%	317×10^3	21 mg/dL	0.9 mg/dL

Source: Hillyer, C.D. et al., *J. Med. Primatol.*, 24, 68, 1995.
Notes: n = 4 HUS monkeys; n = 2 control monkeys.

VITAMIN E DEFICIENCY

A documented, well-characterized anemia is generated in rhesus monkeys that are maintained on vitamin E–deficient diets over extended periods (typically 1.5–2 years) (Dinning and Day 1957; Porter et al. 1962; Fitch et al. 1980). The features of this anemia other than a reduction of the erythrocyte count, hematocrit, and hemoglobin include a shortened erythrocyte life span, a modest reticulocyte response, and some unusual developmental erythromorphologic presentations. In regard to the latter, many multinucleated erythroid precursors are identified in the marrow. All stages of maturation are involved with cells having up to and including four nuclei. The most common aberrant cells are binucleated. In one study of vitamin E–deficient *Macaca mulatta* (n = 5), the proportion of multinucleated erythroid precursors ranged from 12% to 34%, mean 18% of marrow hemopoietic cells (Porter et al. 1962). The nuclei of all the erythrocytic precursors tended to have a homogeneous densely staining appearance and lacked the typical clumped chromatin of normoblasts. The staining was particularly dark and intense in the periphery of the nuclei (Figure 43). The size of the cells varied directly with the number of nuclei. The increment in dimensions of a given cell appeared to be due to the nuclear mass and not to an increase of cytoplasm. Those cells with one nucleus were normal in size and those with four nuclei were quite large. The size of the nuclei was commensurate with the degree of cytoplasmic maturation. The cytoplasm of the cells was normal. The morphologic pattern was not megaloblastic. Another pertinent observation of the bone marrow is the occurrence of erythroid hyperplasia. In the investigation of Porter et al., normoblasts totaled a mean 39% of the hemopoietic cells of the bone marrow in the anemic subjects versus 25% in the normal control animals (Table 8). The erythrocytes in circulation are normocytic and normochromic, but consistent with the pattern observed in the marrow, a few can be binucleated. Of note is the fact that the reticulocyte response is inappropriately modest in the range of 1.4%–3.2% (Porter et al. 1962; Fitch et al. 1980). The osmotic fragility after 24 hours of incubation at 37°C was within normal limits while the erythrocyte ^{51}Cr survival was significantly shortened (10.1 days versus 13.6 days for normal subjects) (Porter et al. 1962; Fitch et al. 1980). The combined marrow and blood profile characterizes this anemia as one in which ineffective erythropoiesis significantly contributes to the erythroid deficit. Ineffective erythropoiesis is corroborated by the decreased utilization of radioiron for the production of circulating hemoglobin in these monkeys (Fitch et al. 1980). The verified abbreviation of the erythrocyte life span also indicates a hemolytic component. Following treatment with D-α-tocopherol acetate, the hematologic and biochemical alterations of the anemic monkeys return to normal.

Cynomolgus monkeys similarly develop an anemia when maintained on a vitamin E–deficient diet (Table 1, Santiyanont et al. 1977; Boonjawat et al. 1979). A reticulocyte response does not develop, and the circulating erythrocytes are normocytic. The anemia appears after at least 15 months of deprivation of the vitamin. Unlike the erythrocytes of the vitamin E–deficient rhesus monkey which do not demonstrate any changes in their response to osmotic challenge (Porter et al. 1962),

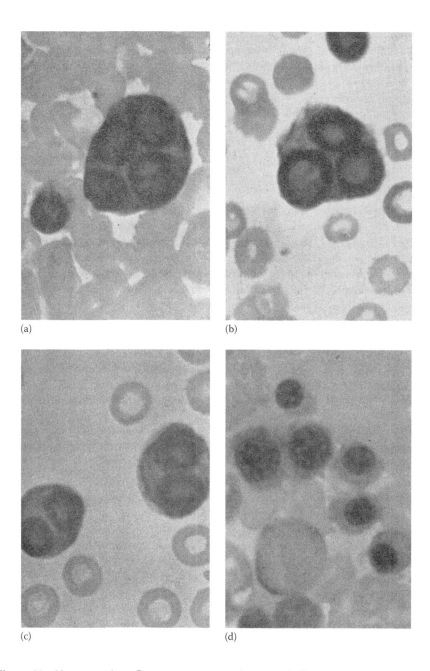

(a)   (b)

(c)   (d)

Figure 43 *Macaca mulatta.* Bone marrow smears from vitamin E–deficient rhesus monkeys. (a–c) Illustrate multinucleated erythroid precursors. The nuclei have a homogeneous appearance and lack the typical clumping of the chromatin that is characteristic of the normoblastic nucleus. The staining is dark and intense along the periphery of the nuclei. (d) Presents normal normoblasts from a monkey during recovery from vitamin E deficiency. (From Porter, F.S. et al., *Blood*, 20, 471, 1962.)

the Rbc of vitamin E–deficient *Macaca fascicularis* is more resistant to osmotic lysis (Santiyanont et al. 1977). Bone marrow was aspirated from the posterior iliac crest of the latter workers' vitamin-deficient and control monkeys and morphologically analyzed (Table 15). The anemic monkeys demonstrated erythroid hyperplasia with an accumulation of orthochromatophilic normoblasts and the presence of multinucleated erythroid precursors with densely stained nuclei. The limited studies (i.e., small numbers of experimental vitamin-deficient rhesus and cynomolgus monkeys) of Porter and of Santiyanont suggest that the multinucleated normoblasts may be less numerous in the long-tailed macaque. The addition of vitamin E to the diet results in a reticulocytosis and a return of the normal morphology of the bone marrow (Table 15).

A macrocytic hemolytic anemia is generated in cynomolgus monkeys when they are maintained on an experimental vitamin E–deficient diet containing 8% stripped safflower oil (polyunsaturated oil) (22% of the calories) for a period of 2 years (Ausman and Hayes 1974, Hayes 1974). The studied monkeys were laboratory raised and were placed on the diet at 1 year of age. After 2 years on the diet, a macrocytic hemolytic anemia appeared and was accompanied in the late stages with slight jaundice, muscular weakness, and severe weight loss. A reticulocytosis was identifiable. Of particular erythromorphologic interest was the occurrence of megaloblastic changes in the bone marrow in the terminal stages of the disease along with erythroid hypercellularity and multinucleated red cell precursors. The megaloblastosis, it was theorized, could have been due to localized folate deficiency. Hemosiderosis attributed to hemolysis was identified in the liver, spleen, and lungs. The anemia does not occur if coconut oil is used instead of safflower oil.

EXPERIMENTALLY INDUCED AUTOIMMUNE HEMOLYTIC ANEMIA

An autoimmune hemolytic anemia is an anemia in which erythrocytes are destroyed by autoantibodies. This condition has three major etiologies, that is, primary (idiopathic), secondary to conditions such as malignancies (e.g., lymphoma) and collagen diseases, and those induced by drugs. Autoimmune hemolytic anemia (AIHA) has been infrequently identified in nonhuman primates. In the present circumstance, three infant *Macaca mulatta* were artificially nursed from the age of 10 days onward for the purpose of neurophysiologic experimentation. They were all administered four drugs in multiple doses throughout the interval of 88–368 days of age. The drugs were bromocriptine methanesulfonate (a dopamine agonist), bicuculline methiodide (γ-aminobutyric acid-A receptor inhibitor), phaclofen (γ-aminobutyric acid-B receptor inhibitor), and SCH 23390 hydrochloride (dopamine-D1 receptor inhibitor) and were injected into the neocortex of the infants' brains (Suzuki et al. 2000). On day 234 of the experiment, it was noted that one of the subjects demonstrated a cessation of weight gain, diarrhea, and pallor. The first hematologic examination was performed on this day and revealed that the subject had a striking anemia (Rbc $91 \times 10^4/\mu L$, Hct 8.2%, and Hb 2.5 g/dL). Other pertinent findings were the presence of normoblasts in the circulation, and

Table 15 Bone Marrow Erythroid Differential Counts of *Macaca fascicularis*

	Site	Age	Sex, n =	ProNbl	BasNbl	Polych	Orth	Mean Nbl	Range
Santiyanont[a] et al. (1977)									
Normal monkeys	Ilium	Young	?, 3	3.0	6.2	24.9	65.8	—	—
Vitamin E–deficient	Ilium	Young	?, 3	3.7	1.6	10.3	86.5	—	—
Multinucleated erythroid cells: 2.8									
Vitamin E–deficient/ supplemented	Ilium	Young	?, 1	8.8	—	16.2	75.0	—	—
No multinucleated erythroid cells									
Winkle[b] (1951)									
Normal monkeys	Ilium	10–15 mo	♂, 2	0.9	3.6	25.6	6.25	36.3	27.4–45.2

Notes: Values are mean values. ProNbl, percentage of pronormoblasts; BasNbl, percentage of basophilic normoblasts; Polych, percentage of polychromatophilic normoblasts; Orth, percentage of orthochromatophilic/orthochromic normoblasts.
[a] Values for Santiyanont are relative percentages of normoblasts of different levels of maturation. That is, the total percentage of normoblasts equals 100%. Vitamin E–deficient monkeys have multinucleated normoblasts in their bone marrow. One vitamin E–deficient subject was supplemented with DL–α–tocopherol for 10 days and then subjected to a bone marrow aspiration
[b] Values for Winkle are mean quantities of normoblasts expressed as the percent of the total population of nucleated mature and immature hemopoietic cells observed in Romanowsky-type dye-stained dry film smears of marrow. Range is the range of the total percentage of normoblasts identified in the marrow.

palpable splenomegaly. Direct Coomb's tests (direct antiglobulin tests, DAT) on that date and later were positive indicating that the surfaces of the red cells were coated with autoantibodies. An incidental observation was that the MCV derived from the initial erythrogram was high for the species, 90 fL, indicating a regenerative macrocytosis, reticulocytosis, and probable erythroblastic hyperplasia secondary to the hemolysis caused by the autoantibodies. The concentration of iron in the serum was ~400 μg/dL, that is, two to three times higher than the average in normal animals. The anemic monkey was treated with steroids and transfusions of erythrocytes in order to sustain it and permit the completion of the neurophysiologic studies. The subject was euthanized and at autopsy a prominent splenomegaly and extramedullary hemopoiesis in the liver and kidneys were observed. The agent or combination of agents responsible for the development of the autoantibodies was not identified.

PARVOVIRUS INFECTION

The Parvoviridae are a family of single-stranded DNA viruses that are the smallest known viruses that infect mammalian cells. (Note: *parvus* is the Latin adjective for small.) A key feature of the Parvoviridae is their requirement for actively dividing cells in order to replicate. The human B19 parvovirus, a constituent of this family, is an important pathogen in man. Based on epidemiologic studies, it has a seroprevalence of ~60% in the human population. This virus is recognized for its infection of human fetuses and human bone marrow. Its infection of erythropoietic cells results in the development of a severe anemia.

The isolation of parvovirus from cynomolgus monkeys with a marked anemia has been reported by O'Sullivan et al. (1994). These monkeys were a cluster of five subjects that were members of colony of ~1200 cynomolgus monkeys at the Comparative Medicine Clinical Research Center of the Bowman Gray School of Medicine. A 723-base-pair fragment of viral DNA that has a 65% homology of human B19 parvovirus has been isolated from the serum of these anemic long-tailed monkeys. Microscopic examination of bone marrow from the proximal femur or sternebrae of the subjects revealed a moderate to marked loss of the further developed erythroid (and myeloid) progenitors along with the presence of many to medium to large undifferentiated cells. The morphologic erythrocellular presentation was variable. A frequent finding in bone marrows having a moderate diminution of erythroid precursors revealed marked dyserythropoiesis. Noticeably abnormal cells of erythroid lineage presented bizarre nuclei, nuclear blebs, nuclear appendages, and multilobulation. Light microscopy showed intranuclear inclusion bodies in the normoblasts and ultrastructural analyses revealed viral arrays characteristic of parvoviruses. The anemia could be classified as nonregenerative on the basis that it was severe but did not generate a reticulocyte response. An unusual feature concerning the severe anemia that is engendered by this virus is that it is consistently associated with the presence of a known immunosuppressive virus, type D simian retrovirus. Further, it has been speculated that type D simian retrovirus may predispose the animals to infection with simian parvovirus and perhaps even be a prerequisite for the development of anemia in animals infected with parvovirus (O'Sullivan et al. 1994).

The study of a second group of cynomolgus monkeys infected with simian parvovirus at a separate facility conducted by the same group of investigators cited earlier has led to the confirmation of some of the original conclusions concerning this virus's activity as well as further insight into its pathology (O'Sullivan et al. 1996). It has been consequently hypothesized that infection of monkeys with simian parvovirus is often inapparent or has mild consequences and that severe illness, such as anemia, ensues only in the presence of predisposing factors such as immunodeficiency initiated by concurrent infection with simian retrovirus or perhaps certain drugs that permit a persistent infection with ongoing destruction of erythroid precursors (O'Sullivan et al. 1996). This situation leads to a chronic destruction of normoblasts and over time results in a persistent nonregenerative anemia.

The workers point out that an experimental infection of monkeys with simian parvovirus is transient and mild suggesting the possibility that a given individual severely anemic monkey with whose anemia is ascribed to a documented simian retrovirus infection may also in fact have an unrecognized concurrent simian parvovirus infection.

It appears that the anemia induced by parvovirus may be predominantly nonregenerative or regenerative. The anemic condition may present a reticulocytosis early in the process and then progress to the nonregenerative stage. The bone marrow can initially mount a regenerative response and then lose this ability in the face of continuous destruction of the developing normoblasts.

It has been proposed that cynomolgus monkeys infected with this virus may serve as an animal model for human infections of the B19 parvovirus.

PYRIDOXINE DEFICIENCY

Macaca mulatta monkeys (n = 16) were placed on a pyridoxine (vitamin B_6)-deficient diet for periods of several months to over 1 year (Poppen et al. 1952). An anemia invariably developed by the second month following the withdrawal of the vitamin and persisted until the animal expired or was given a pyridoxine supplement. The anemic monkeys developed a poor appetite by the second or third week, eventually accompanied by weight loss, unkempt appearance, hair loss with graying, loosening of teeth, orbital edema, apathy or hyperirritability, and hepatic biochemical changes.

The blood picture presented a decrease in the Rbc count and hemoglobin that continued to diminish until at least the 25th week of the deficient diet. Between the 35th and 45th week, there was a transient rise in the erythrocyte count and hemoglobin concentration without any concomitant clinical improvement and which did not materially alter the progress of the anemia. This interval was interpreted as a temporary period of dehydration and hemoconcentration. All the erythrocyte indices (MCV, MCH, MCHC) decreased during the course of the disease, and the mean cellular diameter increased from ~7.0 to ~8 μm. These alterations resulted in the appearance of macrocytic hypochromic erythrocytes in circulation. Calculations indicated that these red cells were markedly thinner than normal. At the height of the anemia, a marked anisocytosis, slight poikilocytosis, polychromasia, and a circulating normoblastosis that paralleled the development of the anemia comprised the morphologic profile. The deficiency in the production of hemoglobin was correlated with pyridoxine's role in protein metabolism as well as likely associated with the hepatic abnormalities that were identified at necropsy (fatty infiltration and cirrhosis). Normoblastic hyperplasia of the sternal or tibial marrow was regularly noted and in occasional instances brought about a reversal of the myeloid–erythroid ratio (M:E ratio, i.e., the ratio of the quantity of mature and immature neutrophils to the quantity of developing [nucleated] erythroid cells in the bone marrow). The thin hypochromic red cells produced in pyridoxine deficiency, as might be expected, demonstrate an increased resistance to osmotic hemolysis. This pool of monkeys' erythrocytes demonstrated complete hemolysis when exposed to 0.24% sodium chloride solution while normal monkeys' red cells underwent complete hemolysis in 0.33% saline.

In another earlier study of rhesus monkeys placed on a pyridoxine-deficient diet (n = 4 young, immature, ~1.5–2.0 kg), the monkeys failed to gain weight and grow normally, and presented physical features similar to those attributed to the above-described vitamin-deficient subjects (McCall et al. 1946). The blood picture included hypochromic microcytic anemia, polychromasia, and normoblastosis.

CHAPTER 25

Osmotic Fragility

The osmotic fragility test of red cells is essentially a measurement of the extent of redundancy of the erythrocytes' cell wall, or as expressed by Beutler (1977), it is a simplified means of estimating the surface–volume ratio of erythrocytes. That is, the more redundant a cell's wall is, the greater is its capability to accommodate to hypotonic solutions. As one might conclude, biconcave-shaped cells (such as the macaque's or man's erythrocytes) have a greater redundancy of plasmalemma (an excess of plasma membrane) than spherocytic cells (Figure 28).

When the osmotic fragility test is carried out, the erythrocytes are placed in a hypotonic medium (sodium chloride solution), and the red cells fill with water until the osmotic pressure inside the cell is reduced to that of the surrounding medium. As long as the intracellular osmotic pressure is greater than the solution in which it is immersed, the challenged red cell will continue to imbibe water until it becomes a perfect sphere. Once the erythrocyte reaches this configuration, no membrane is available for further expansion of the cell and if equilibrium between intracellular and extracellular osmotic pressures has not been attained at this juncture, water will continue to enter the cell, and this results in rupture of the erythrocyte. In the case of man's red cells, for example, it becomes a sphere when it reaches a volume that is 1.8 times its resting volume.

The fragility test is conducted by placing the test cells in a series of 10 saline solutions that are increasingly hypotonic. The amount of hemolysis derived at each concentration is determined by quantitating the amount of hemoglobin that has been liberated into solution by the hemolyzed cells. Of course, concentrations that are accommodated by the red cells do not yield any lysis and liberation of hemoglobin. Once the hypotonic challenge is sufficient to lyse at least some erythrocytes, this point is recognized as "initial hemolysis" and the following increments in hypotonicity will result in increasing amounts of hemolysis until a given concentration causes a hemolysis of all cells in a sample. Under normal conditions, man's red cells do not lyse until the saline concentration is 0.45% or less. Hemolysis of 50% of the erythrocytes occurs with ~0.40% saline and 100% lysis is attained with exposure to ~0.35% saline (Kjeldsberg et al. 1989). It should be noted that the term *lysis* as it is used in the context of osmotic lysis is inexact as the membrane alteration is reversible and cells that have undergone osmotic hemolysis can regain their osmotic integrity (Wintrobe et al. 1974).

Krumbhaar and Musser (1921) may have been the first to study the "resistance of erythrocytes" of *Macacus rhesus* to hypotonic saline solutions. They studied 13 normal monkeys of undetermined age with a weight range of 2.05–6.02 kg. In their experience, mean initial hemolysis was noted at 0.46% saline and complete hemolysis was obtained with exposure to the 0.33% solution. The most resistant cells of any animals showed beginning and total hemolysis at 0.44% and 0.30%, respectively. The monkey with the least resistant cells presented beginning hemolysis at 0.48% saline and complete hemolysis with 0.36% salt solution. Wills and Stewart (1935) derived the same hypotonic sensitivity for their immature rhesus monkeys. They sampled some of their 13 normal subjects and observed that the red cells began hemolysis at 0.45% saline and completed it at 0.35%. It is seen that the osmotic fragility of human and the rhesus monkeys' Rbc are comparable even though the monkeys' red cells are considerably smaller (Wintrobe et al. 1974; Beutler 1977; Kjeldsberg et al. 1989).

Ponder et al. (1928b, 1929) reported that the erythrocytes of *Macaca mulatta* were more resistant to hypotonic hemolysis than man's red cells whose erythrocytes they stated were able to resist 0.32% NaCl. They cited the Rbc of *Macaca mulatta* as being able "to resist" 0.27% saline while the Rbc of the cynomolgus monkey were even more resistant to hypotonic challenge. They were resistant to 0.17% saline.

When a sample of blood is incubated for 24 hr at 37°C before the osmotic fragility test is performed, the test erythrocytes show increased sensitivity to osmotic lysis. That is, hemolysis initiates at a higher concentration of saline and hemolysis of 50% of the red cells similarly occurs at a higher concentration of saline than observed with red cells that have not undergone incubation in their own serum prior to analysis. In man according to the data of Beutler (1977, Kjeldsberg et al. 1989), man's incubated normal red cells present 10% hemolysis at 0.57%–0.65% NaCl and 50% hemolysis when incubated in 0.49%–0.59% saline. In the case of *Macaca mulatta*, the incubated red cells display 10% hemolysis in hypotonic 0.67%–0.38% NaCl and 50% hemolysis in 0.39%–0.21% saline (n = 10, Figure 44, Porter et al. 1962). It thus appears from this single comparison that autologous serum-incubated red cells of the rhesus monkey may initially hemolyze sooner than man's incubated red cells, but some individuals withstand a greater hypotonic challenge than the normal human red cells analyzed by Beutler. And in an extension of this trend, 50% hemolysis of monkeys' red cells required greater hypotonicity than required for human Rbc to demonstrate this level of hemolysis. That is, the monkeys' cells were more resilient to osmotic challenge than man's Rbc. This situation is not readily understood and seems counterintuitive to the concept that smaller red cells are typically more sensitive to hypotonic challenge than larger erythrocytes. It could be assumed that the different osmotic sensitivities of rhesus and man's Rbc following a prior 24 hr incubation is dependent on the net difference of the two species metabolic effects imposed upon the erythrocytes during incubation (e.g., rate of utilization of plasma glucose or accumulation of certain metabolites).

Krumbhaar and Musser's (1923) early studies of osmotic fragility of rhesus monkeys' erythrocytes included osmotic fragility of erythrocytes circulating in the blood after splenectomy. Six splenectomized subjects and two controls were monitored

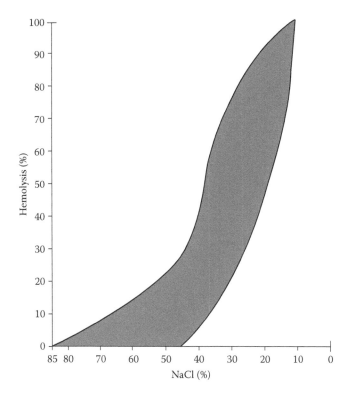

Figure 44 *Macaca mulatta*. Osmotic fragility of the erythrocytes of rhesus monkeys (*Macaca mulatta*). The red cells were incubated in their own serum at 37°C for 24 hours and then placed in a series of hypotonic solutions for 1 hour. The degree of hemolysis was determined by quantitation of the hemoglobin released by the hemolyzed red cells. Values on the left side of the graph indicate the proportion of red cells (in percentage) that underwent complete lysis when exposed to a given level of salinity. The darkend area presents the range of hemolysis derived from a composite of 10 different monkeys' fragility curves. The concentrations of NaCl as labeled in the graph should have a decimal point prior to the digits, that is, .85% NaCl, .80% NaCl, and so forth. (From Porter, F.S. et al., *Blood*, 20, 471, 1962.)

for periods of 1–2 years post-splenectomy (Figure 45). It was noted that a definite mean increase in resistance of initial hemolysis was identifiable. It presented at 0.42% saline (control = 0.45%). And complete hemolysis was achieved with 0.26% sodium chloride as opposed to 0.33% for the non-splenectomized controls. Each splenectomized subject demonstrated this change, and this modification of response to osmotic challenge persisted through the 2 years of investigation. This increased resistance to hypotonic challenge could not be directly attributed to a postoperative anemia. The subjects generally demonstrated a transient post-splenectomy anemia, but on average it persisted for only 1 month. The cause of the increased resistance to hypo-osmotic challenge was not identified but the data do suggest that the osmotic characteristics of the population of erythrocytes in circulation following splenectomy are slightly different from the presurgical population. It can be postulated that the

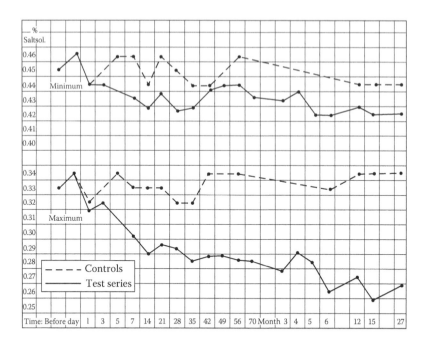

Figure 45 *Macaca mulatta*. Resistance of erythrocytes of splenectomized rhesus monkeys to lysis induced by hypotonic saline. The top pair of curves indicate the concentrations of saline required for the occurrence of initial hemolysis during a period of prior to splenectomy, and thereafter after splenectomy for the following 27 months. The bottom pair of curves indicate the concentrations at which complete hemolysis is attained during the same interval. The curves are composite curves for six splenectomized monkeys (solid lines) and two control subjects (dotted lines). It is evident that splenectomy is associated with an increase in resistance of the erythrocytes to hemolysis caused by hypotonic saline both for initial hemolysis and particularly for complete hemolysis. The alteration in susceptibility to hypotonic challenge is observed to persist for at least 27 months post-splenectomy. (From Krumbhaar, E.B. and Musser, Jr., J.H., *Arch. Int. Med.*, 31, 686, 1923.)

cell surface molecular signals proffered by aged and damaged red blood cells engender a slightly different response by the phagocytic cells of the bone marrow and liver than when the spleen is the major site of red cell destruction. The net result may well be a greater rate of cell loss and a consequent compensatory increased production and release of immature red cells into circulation. These presumed reticulocytes would be expected to be minimally larger than mature erythrocytes and because of their enhanced size be slightly more resistant to osmotic hemolysis. This subset could modify the osmotic profile of the population of circulating erythrocytes. This concept could be further investigated by confirming the experimental results of Krumbhaar and Musser coupled with establishing the relative and absolute reticulocyte counts of splenectomized monkeys (along with controls) along with the quantitation of ribonucleoprotein in the reticulocytes of the test and control subjects.

CHAPTER 26

ABO Blood Group System

Rhesus and cynomolgus monkeys, unlike humans, do not have AB cell surface antigens on the surfaces of their erythrocytes. They do, however, have anti-A and anti-B agglutinins in their sera (Sae-Low and Malaivijitnond 2003). In view of the current interest in stem cell and transplantation studies in monkeys, recognition of the blood types of both donors and recipients has become important. Chi et al. (2010) have developed a flow cytometric system to detect ABO blood group antibody levels in rhesus and cynomolgus monkeys. In this system, human A and B red blood cells are used as target cells. The binding of the target cells with anti-A or anti-B blood group antibodies in sera of rhesus and cynomolgus monkeys is detected by flow cytometry after adding secondary specific fluorescence-labeled antihuman IgG or IgM antibody. The sera have to be preabsorbed on normal human type O Rbc to remove nonspecific human antibodies. This method yields an accurate detection of ABO antibodies, and it has been found that the concentrations of anti-A and anti-B antibodies are significantly lower in monkey sera than they are in human serum.

A method to detect the A, B, and AB phenotypes of rhesus and cynomolgus macaques based on real-time quantitative polymerase chain reactions (PCRs) has also been developed by Premasuthan et al. (2012). In addition to the method's high order of specificity it has the advantage of eliminating the need for fresh blood (or saliva). The latter group reports that the distribution of the A, B, and AB phenotypes of the studied cynomolgus macaques while regionally variable resembles that of the rhesus monkey and that the blood group B is predominant in both species. Of the 89 studied cynomolgus monkeys, 18 were of type A, 43 were of type B, and 28 were of type AB. As for the 78 rhesus monkeys that were tested, 16 were found to be of type A, 40 were of type B, and 22 were of type AB.

Although it is usually accepted that *Macaca fascicularis* monkeys do not present human ABO blood groups on the surface of their erythrocytes, the work of Qiong et al. (2003) suggests that this model may not be invariably true. Small amounts of these antigens may well be identifiable on the surface of red cells of a segment of the cynomolgus monkey population. In a study of 30 *Macaca fascicularis*, all subjects uniformly failed to demonstrate the red cell surface antigens by direct hemagglutination testing. However, the ABO blood groups were identified on the erythrocytes of certain monkeys by the heat absorption elution test. In that circumstance, 3 monkeys offered cells with the type A antigen, 9 displayed the B red cell surface antigen, 3 had both A and B markers, and the remaining 15 monkeys presented neither A nor B antigens.

Table 1 Erythrocyte Counts and Related Values

	RBC	Hct	Hb	Diameter	MCV	MCH	MCHC
Phylum Chordata							
Subphylum Vertebrata							
Class Mammalia							
Order Primates							
Family Cercopithecidae, old world monkeys							
Genus *Macaca*							
Macaca mulatta, rhesus monkey/macaque							
Macaca mulatta[1], rhesus monkey/macaque, captive colony, ketamine							
Macaca mulatta[1]♂, adult, 45–192 mo, mean 10,443 g, n = 27	5.06	37.6	12.8	—	74	25	34
Macaca mulatta[1]♀, adult, 45–192 mo, mean 8,575 g, n = 30	5.08	36.7	12.5	—	73	25	34
Macaca mulatta[2]♀, scheduled analyses over 3–33 months, mean 19 mo							
Macaca mulatta[2]♀, captive, adult, n = 15, 107 observations	5.19	41.1	13.4	—	79	26	33
Macaca mulatta[3]♂, captive, young adult, mean 4.3 kg, n = 42	4.65*	40.9*	12.5	—	90	27	31
Macaca mulatta[3]♀, captive, young adult, mean 3.7 kg, n = 60	4.35*	39.9*	12.4	—	94	29	31
Macaca mulatta[4]♂, captive, adult, mean 5.8 yr old, n = 5	5.71	43.8	13.8	—	77	25	32
Macaca mulatta[4]♀, captive, adult, mean 8.2 yr old n = 20	5.79	44.6	14.2	—	78	25	32
Macaca mulatta[5a]♂, ♀, captive, housed singly, 4.5–6 kg, sampling before A.M. feeding							
Macaca mulatta[5a]♂, 3.5–6 yrs age, assay weekly x40, n = 5	5.54*	41.3*	13.2*	—	75	24	32
Macaca mulatta[5a]♀, 3.5–6 yrs age, assay weekly x40, n = 6	5.10*	38.7*	12.3*	—	76	24	32
Macaca mulatta[5b]♂, ♀, captive, housed singly, sampling before A.M. feeding							
Macaca mulatta[5b]♂, <3 yr, assay every 2 wk in 12 weeks, n = 36	5.70ˣ	41.9*	14.1*ˣ	—	74	25	34
Macaca mulatta[5b]♀, <3 yr, assay every 2 wk in 12 weeks, n = 22	5.57	40.9*	13.4*	—	73	24	33

(*Continued*)

185

Table 1 (Continued) Erythrocyte Counts and Related Values

	RBC	Hct	Hb	Diameter	MCV	MCH	MCHC
Macaca mulatta[5c]♂, >3 yr, assay every 2 wk in 12 weeks, n = 38	5.85*x	42.1*	13.8*x	—	72	24	33
Macaca mulatta[5c]♀, >3 yr, assay every 2 wk in 12 weeks, n = 34	5.60*	41.1*	13.5*	—	74	24	33
Macaca mulatta[8], captive, adult, 7–8 yrs old, before A.M. feeding							
Macaca mulatta[8]♂, captive, adult, mean 7.0 kg, n = 6	6.07x	46.6	14.0x	—	77*x	23x	30x
Macaca mulatta[8]♀, captive, adult, mean 4.9 kg, n = 6	6.04y	44.5	13.5y	—	74*y	22y	30y
Macaca mulatta[13], captive-bred, 2.2–3.5 kg, fasted 16 hr, compared with M. fascicularis[13]							
Macaca mulatta[13]♂, captive, approx 3 yr old, n = 26	5.47x	42.0	13.3x	—	77	24	32
Macaca mulatta[13]♀, captive, approx 3 yr old, n = 18	5.40x	42.0x	13.3x	—	78	25	32
Macaca mulatta[16]♂, M. mulatta[16]♀, captive, juvenile, sex immat at the onset of study, approx 2 yr old, two assays/animal, fasted ≥ 18 hr, phlebotomy-accustomed, trained							
Macaca mulatta[16]♂, captive, juvenile, 2.6 kg, n = 5, baseline	5.87	46.0	15.3x	—	78	26	33
Macaca mulatta[16]♀, captive, juvenile, 2.6 kg, n = 5, baseline	5.31	41.0	13.8	—	77	26	34
Macaca mulatta[16]♂, captive, juvenile, 3.1 kg, n = 5, 3rd mo study	5.93	43.0	14.5	—	73	24	34
Macaca mulatta[16]♀, captive, juvenile, 3.2 kg, n = 5, 3rd mo study	5.55	39.0	12.8	—	70	23	33
Macaca mulatta[16]♂, captive, juvenile, 3.4 kg, n = 5, 6th mo study	5.81	45.0	14.7	—	77	25	33
Macaca mulatta[16]♀, captive, juvenile, 3.2 kg, n = 5, 6th mo study	5.55	42.0	14.1	—	76	25	34
Macaca mulatta[16]♂, captive, juvenile, 3.7 kg, n = 5, 9th mo study	5.85	43.0	14.9y	—	74	25	35
Macaca mulatta[16]♀, captive, juvenile, 3.4 kg, n = 5, 9th mo study	5.21	39.0	13.7	—	75	26	35
Macaca mulatta[16]♂, captive, juvenile, 4.1 kg, n = 5, 12th mo study	5.81	43.0	13.3xy	—	74	23	31
Macaca mulatta[16]♀, captive, juvenile, 3.9 kg, n = 5, 12th mo study	5.19	39.0	12.9	—	75	25	33
Macaca mulatta[16]♂, captive, juvenile, sex immat, mean 2.64 kg, n = 25 (50 analyses), subjects of this study fasted ≥ 18 hr, phlebotomy-accustomed, trained							
Macaca mulatta[16]♂, captive, juvenile, sex immat, n = 25	5.79	43.9a	14.7	—	76	25	33
Macaca mulatta[16]♀, captive, juvenile, sex immat, mean 2.68 kg, n = 23 (46 analyses)				—			

(Continued)

TABLE 1   187

Table 1 (Continued)   Erythrocyte Counts and Related Values

	RBC	Hct	Hb	Diameter	MCV	MCH	MCHC
Macaca mulatta[16]♀, captive, juvenile, sex immat, n = 23	5.51	43.3	14.6	—	79	26	34
Macaca mulatta[16]♂, captive, adult, sex mature, mean 8.86 kg, n = 17 (78 analyses)				—			
Macaca mulatta[16]♂, captive, adult, sex mature, n = 17	5.85	46.6[a]	15.2	—	80	26	33
Macaca mulatta[16]♀, captive, adult, sex mature, mean 7.00 kg, n = 15 (76 analyses)				—			
Macaca mulatta[16]♀, captive, adult, sex mature, n = 15	5.64	44.3	14.8		79	26	33
Macaca mulatta[16]♂ and ♀, longitudinal study, mature monkeys, two assays/animal/interval, subjects of this study fasted ≥ 18 hr, phlebotomy-accustomed, trained							
Macaca mulatta[16]♂, captive, mean 8.46 kg, n = 4, baseline	5.58[b]	45.0	14.6[b]	—	81	26	32
Macaca mulatta[16]♀, captive, mean 6.50 kg, n = 4, baseline	5.54	44.0	14.8	—	79	27	34
Macaca mulatta[16]♂, captive, mean 9.48 kg, n = 4, 6th mo study	6.10	50.0	17.2[b]	—	82	28	34
Macaca mulatta[16]♀, captive, mean 7.80 kg, n = 4, 6th mo study	5.96	48.0	16.7	—	81	28	35
Macaca mulatta[16]♂, captive, mean 10.1 kg, n = 4, 12th mo study	6.46[b]	51.0	16.6	—	79	26	33
Macaca mulatta[16]♀, captive, mean 7.96 kg, n = 4, 12th mo study	5.79	47.0	16.4	—	81	28	35
Macaca mulatta[16]♂, captive, mean 10.1 kg, n = 4, 18th mo study	6.20	48.0	14.6	—	77	24	30
Macaca mulatta[16]♀, captive, mean 8.46 kg, n = 4, 18th mo study	5.80	44.0	14.3	—	76	25	33
Macaca mulatta[16]♂, captive, mean 9.79 kg, n = 4, 24th mo study	5.89	47.0	15.3	—	80	26	33
Macaca mulatta[16]♀, captive, mean 8.40 kg, n = 4, 24th mo study	6.02	46.0	15.0	—	76	25	33
Macaca mulatta[16]♂, captive, mean 10.8 kg, n = 4, 30th mo study	6.05	50.0	16.2	—	83	27	32
Macaca mulatta[16]♀, captive, mean 9.04 kg, n = 4, 30th mo study	5.98	46.0	14.8	—	77	25	32
Macaca mulatta[16]♂, captive, mean 10.6 kg, n = 4, 36th mo study	5.60	44.0	15.3	—	79	27	35
Macaca mulatta[16]♀, captive, mean 9.00 kg, n = 4, 36th mo study	5.55	44.0	15.1	—	79	27	34
Macaca mulatta[16]♂, captive, n = 4, 3 yr mean	5.98	47.9	15.7	—	80	26	33
Macaca mulatta[16]♀, captive, n = 4, 3 yr mean	5.81	45.6	15.3	—	78	26	34

(*Continued*)

Table 1 (Continued) Erythrocyte Counts and Related Values

	RBC	Hct	Hb	Diameter	MCV	MCH	MCHC
Macaca mulatta[19], 9♂, 9♀, captive, 3–4.5 kg, each tested with each medication IM							
Macaca mulatta[19], 9♂, 9♀, 3 yr old, ketamine	5.16	36.4	11.5	—	71	22	32
Macaca mulatta[19], 9♂, 9♀, 3 yr, ketamine and acepromazine	5.26	37.1	11.5	—	71	22	31
Macaca mulatta[19], 9♂, 9♀, 3 yr, tiletamine and zolazepam	5.17	36.7	11.4	—	71	22	31
Macaca mulatta[23], 18♂, initial 2.9 final 5.1 kg; 18♀, initial 2.8 final 4.6 kg, >4 yr of age at the end of study daily contact per gastric supplementation of food while in restraining box							
Macaca mulatta[23], 18♂, 18♀, young captive, initial study	4.32	—	13.4	—	—	31	—
Macaca mulatta[23], 18♂, 18♀, young captive, 3rd mo	3.96	—	11.6	—	—	29	—
Macaca mulatta[23], 18♂, 18♀, young captive, 6th mo	3.94	—	12.4	—	—	31	—
Macaca mulatta[23], 18♂, 18♀, young captive, 9th mo	4.25	—	11.9	—	—	28	—
Macaca mulatta[23], 18♂, 18♀, young captive, 12th mo	3.74	—	10.8	—	—	29	—
Macaca mulatta[24]♀, Phase I, free living then capture, day 1, 22–25 mo, 2.1–4.8 kg, mean 3.6 kg							
Macaca mulatta[24]♀, capture, period acute stress, ketamine, n = 45	4.63	35.8	10.7	—	77	23	30
Macaca mulatta[24]♀, Phase II, single housing, presumed chronic stress, ketamine							
Macaca mulatta[24]♀, first assay, 6 mo study, n = 45	5.55	41.3	12.5	—	76	23	30
Macaca mulatta[24]♀, second assay, 6 mo study, n = 45	5.49	41.1	12.5	—	75	23	30
Macaca mulatta[24]♀, third assay, 6 mo study, n = 45	5.18	41.0	12.0	—	79	23	29
Macaca mulatta[24]♀, fourth assay, 6 mo study, n = 45	5.32	42.3	12.1	—	79	23	29
Macaca mulatta[24]♀, fifth assay, 6 mo study, n = 45	5.33	40.8	12.7	—	77	24	31
Macaca mulatta[24]♀, mean 6 mo study period, n = 45	5.37	41.3	12.3	—	77	23	30
Macaca mulatta[24]♀, Phase III, social housing, normalized environment, ketamine, n = 45							
Macaca mulatta[24]♀, social housing, 3 mo post entry, n = 45	5.40	41.5	12.7	—	77	23	30
Macaca mulatta[24]♀, social housing, 7 mo post entry, n = 45	5.31	41.0	12.4	—	77	23	30
Macaca mulatta[24]♀, social housing, mean, 7 mo period, n = 45	5.36	41.3	12.5	—	77	23	30

(Continued)

TABLE 1   189

Table 1 (Continued) Erythrocyte Counts and Related Values

	RBC	Hct	Hb	Diameter	MCV	MCH	MCHC
Macaca mulatta[25], captive, aged, 15–28 yr old, ♂ mean 8.9 kg, ♀ mean 7.9 kg, fasted overnight with access to water, ketamine							
Macaca mulatta[25]♂, captive, aged, 20.3 yr mean, n = 7	6.30	46.7	14.3	—	75x	23	30
Macaca mulatta[25]♀, captive, aged, 19.6 yr mean, n = 17	5.90	42.0x	12.5	—	72	22x	30
Macaca mulatta[25], captive, adult, 6–14 yr old, mean 7.5 yr ♂, mean 9.4 kg; 4–10 yr old mean 7.4 yr ♀, mean 7.2 kg, ketamine, fasted overnight with access to water							
Macaca mulatta[25]♂, captive, young adult, 7.5 yr mean, n = 10	6.00	47.8	14.1	—	80x	23	29
Macaca mulatta[25]♀, captive, young adult, 7.4 yr mean, n = 24	5.80	45.4x	13.4	—	80	23x	29
Macaca mulatta[29], colony captive, fentanyl-droperidol IM, fasted 17 hr							
Macaca mulatta[29]♂, 24–54 mo old, 3.4–8.5 kg, n = 56	5.25	42.9	15.1	—	80	29	38
Macaca mulatta[29]♀, 24–80 mo old, 3.1–7.0 kg, n = 57	5.00	41.8	14.6	—	80	27	33
Macaca mulatta[30], free-ranging, age, 2–3 yr old, ♂ mean 3.9 kg, ♀ mean 3.7 kg, capture, placed in single cage, given water, fasted 12 hr							
Macaca mulatta[30]♂, mean 2.3 yr, ketamine, n = 18	5.10*	38.5	12.1	—	76	24	32
Macaca mulatta[30]♀, mean 2.5 yr, ketamine, n = 11	4.20*	32.2	9.8	—	78	24	31
Macaca mulatta[30], free-ranging, age, 4–9 yr old, ♂ mean 10.2 kg, ♀ mean 9.1 kg, capture, placed in single cage, given water, fasted 12 hr							
Macaca mulatta[30]♂, mean 6.2 yr, ketamine, n = 36	5.20	40.4	12.3	—	78	24	31
Macaca mulatta[30]♀, midterm preg, mean 5.8 yr, ketamine, n = 15	5.10	39.9	12.4	—	78	24	31
Macaca mulatta[30], free-ranging, age, ≥ 10 yr old, ♂ mean 10.7 kg, ♀ mean 11.7 kg, capture, placed in single cage, given water, fasted 12 hr							
Macaca mulatta[30]♂, mean 16.2 yr, ketamine, n = 10	5.00	39.7	12.0	—	80	24	30
Macaca mulatta[30]♀, midterm preg, mean 10.8 yr, ketamine, n = 7	4.80	36.5	11.2	—	77	24	31
Macaca mulatta[32], M.D. Anderson Cancer Center Department of Vet Sci Colony							
Macaca mulatta[32]♂, <1 yr, ketamine, n = 27	6.18	41.4	13.1	—	67	21	32
Macaca mulatta[32]♀, <1 yr, ketamine, n = 27	6.04	40.4	12.8	—	67	21	32
Macaca mulatta[32]♂, 1–2 yr, ketamine, n = 30	5.73	40.0	12.8	—	70	22	32
Macaca mulatta[32]♀, 1–2 yr, ketamine, n = 77	5.75	40.1	12.9	—	69	22	32

(Continued)

Table 1 (Continued) Erythrocyte Counts and Related Values

	RBC	Hct	Hb	Diameter	MCV	MCH	MCHC
Macaca mulatta[32]♂, 2–3 yr, ketamine, n = 27	5.71	39.7	13.0	—	68	23	33
Macaca mulatta[32]♀, 2–3 yr, ketamine, n = 50	5.49	39.0	12.6	—	71	23	32
Macaca mulatta[32]♂, 3–4 yr, ketamine, n = 30	5.79	40.2*[a]	12.9*[a]	—	70	22	32
Macaca mulatta[32]♀, 3–4 yr, ketamine, n = 25	5.56	38.6*[a]	12.5*[a]	—	70	23	32
Macaca mulatta[32]♀, 3–4 yr, pregnant, ketamine, n = 11	5.73	40.5	13.0	—	71	23	32
Macaca mulatta[32]♂, 4–5 yr, ketamine, n = 44	5.89	41.1*[a]	13.2*[a]	—	70	22	32
Macaca mulatta[32]♀, 4–5 yr, ketamine, n = 13	5.85	39.5*[a]	12.8*[a]	—	68	22	33
Macaca mulatta[32]♀, 4–5 yr, pregnant, ketamine, n = 20	5.91	41.3	13.3	—	70	23	32
Macaca mulatta[32]♂, 5–10 yr, ketamine, n = 21	5.90	42.4*[a]	13.6*[a]	—	71	23	32
Macaca mulatta[32]♀, 5–10 yr, ketamine, n = 30	5.75	40.3*[a]	12.9*[a]	—	70	22	32
Macaca mulatta[32]♀, 5–10 yr, pregnant, ketamine, n = 44	5.79	40.3	12.9	—	70	22	32
Macaca mulatta[32]♀, >10 yr, ketamine, n = 29	6.01	42.1[a]	13.6[a]	—	70	24	32
Macaca mulatta[32]♀, >10 yr, pregnant, ketamine, n = 22	6.81	40.7	13.5	—	70	23	33
Macaca mulatta[33], originated in southwest China, colony bred in China							
Macaca mulatta[33]♂, 3–5 yr, no anesthesia or fasting, n = 18	5.51*	43.0	13.2*	—	78	24	31
Macaca mulatta[33]♀, 3–5 yr, no anesthesia or fasting, n = 17	5.20*	41.0	12.5*	—	79	24	30
Macaca mulatta[34], captive, semi-fasted, median values							
Macaca mulatta[34]♂, studied over a 6 yr period, 2–6 assays per monkey							
Macaca mulatta[34]♂, av 11 yr age at onset, given ketamine, n = 20	5.84	43.0	13.8	—	74	24	32
Macaca mulatta[34]♀, studied over a 3 yr period, 23–25 assays per monkey							
Macaca mulatta[34]♀, av 11 yr age, captive, bled 3x at onset, not given ketamine, n = 15	5.71	41.0	13.4	—	72	23	32
Macaca mulatta[34], captive infant, no ketamine, median values							
Macaca mulatta[34], 12–24 wk age, 2–3 assays/monkey, n = 9	5.13	38.0	12.3	—	72	24	33

(Continued)

TABLE 1 191

Table 1 (Continued) Erythrocyte Counts and Related Values

	RBC	Hct	Hb	Diameter	MCV	MCH	MCHC
Macaca mulatta[34], captive, ketamine, 42–122 wk age except 46, 50, 58, median values; Macaca mulatta[34], 11–12 assays/infant monkey, n = 9	5.33	38.0	12.5	—	73	24	33
Macaca mulatta[34], captive, no ketamine, 46, 50, 58 wk age, median values; Macaca mulatta[34], 2–4 assays/infant monkey, n = 9	5.66	41.0	13.5	—	72	24	33
Macaca mulatta[34], captive, ketamine, 42, 54, 62 weeks age, mean values; Macaca mulatta[34], 5 males, 4 females, n = 9	5.43[x]	38.0[x]	12.4[x]	—	70[x]	—	—
Macaca mulatta[34], captive, no ketamine, 46, 50, 58 weeks age, mean values; Macaca mulatta[34], 5 males, 4 females, n = 9	5.71[x]	41.0[x]	13.2[x]	—	71[x]	—	—
Macaca mulatta[35]♀, captive, adult, before conception, n = 25	5.57	40.6	12.6	—	73	23	31
Macaca mulatta[35]♀, captive, 0–29 days after conception, n = 25	5.19	39.0	12.5	—	75	24	32
Macaca mulatta[35]♀, captive, 30–59 days after conception, n = 25	5.20	39.4	12.3	—	76	24	31
Macaca mulatta[35]♀, captive, 60–89 days after conception, n = 25	5.12	40.0	126	—	79	25	31
Macaca mulatta[35]♀, captive, 90–99 days after conception, n = 25	5.05	40.0	12.4	—	80	25	31
Macaca mulatta[35]♀, captive, 100–109 days after conception, n = 25	5.02	40.3	12.5	—	80	25	31
Macaca mulatta[35]♀, captive, 110–119 days after conception, n = 25	5.02	39.5	12.4	—	80	25	32
Macaca mulatta[35]♀, captive, 120–129 days after conception, n = 25	4.88	39.1	12.4	—	80	25	32
Macaca mulatta[35]♀, captive, 130–139 days after conception, n = 25	4.75	37.7	11.9	—	80	25	32
Macaca mulatta[35]♀, captive, 140–149 days after conception, n = 25	4.69	37.3	11.5	—	80	25	31
Macaca mulatta[35]♀, captive, 150–159 days after conception, n = 25	4.61	36.5	11.4	—	80	25	31
Macaca mulatta[35]♀, captive, 160–169 days after conception, n = 25	4.44	34.8	11.1	—	80	26	32
Macaca mulatta[35]♀, captive, 170–179 days after conception, n = 25	4.48	35.7	11.1	—	80	25	31
Macaca mulatta[35]♀, captive, 180–189 days after conception, n = 25	4.35	34.5	11.0	—	79	25	32
Macaca mulatta[35]♀, captive, 190–199 days after conception, n = 25	4.50	36.8	11.3	—	82	25	31
Macaca mulatta[35]♀, captive, 200–209 days after conception, n = 25	4.53	36.7	11.5	—	82	26	31

(Continued)

TABLE 1

Table 1 (Continued)   Erythrocyte Counts and Related Values

	RBC	Hct	Hb	Diameter	MCV	MCH	MCHC
Macaca mulatta[36]♀, captive, adult, before conception, n = 42	—	41.5	12.5	—	—	—	30
Macaca mulatta[36]♀, captive, adult, days of pregnancy 1–49, n = 42	—	40.9	12.1	—	—	—	30
Macaca mulatta[36]♀, captive, adult, days pregnant 50–99, n = 42	—	39.4	11.5	—	—	—	29
Macaca mulatta[36]♀, captive, adult, days pregnant 100–109, n = 4	—	39.9	11.9	—	—	—	30
Macaca mulatta[36]♀, captive, adult, days pregnant 110–119, n = 4	—	41.6	11.8	—	—	—	28
Macaca mulatta[36]♀, captive, adult, days pregnant 120–129, n = 42	—	40.2	12.0	—	—	—	30
Macaca mulatta[36]♀, captive, adult, days pregnant 130–139, n = 42	—	38.7	11.6	—	—	—	30
Macaca mulatta[36]♀, captive, adult, days pregnant 140–149, n = 42	—	40.1	11.7	—	—	—	29
Macaca mulatta[36]♀, captive, adult, days pregnant 150–159, n = 42	—	38.2	11.2	—	—	—	29
Macaca mulatta[36]♀, captive, adult, days pregnant 160–169, n = 42	—	38.7	11.4	—	—	—	29
Macaca mulatta[36]♀, captive, adult, days pregnant 170–179, n = 42	—	38.5	11.0	—	—	—	29
Macaca mulatta[31] and M. sinica[31], adult, both sexes, n = 27	—	—	—	6.4	—	—	—
Macaca mulatta[37], plasma/serum	—	—	—	8.0	—	—	—
Macaca mulatta[37], dry film smear	—	—	—	7.3	—	—	—
Macaca mulatta[38], captive, fasted, sampling over several months, n = 8	4.94	—	—	—	—	—	—
Macaca mulatta[39], captive, weanling, iron deficient, n = 27	5.48ˣ	35.5ˣ	11.5ˣ	—	65ˣ	21ˣ	32
Macaca mulatta[39], captive, weanling, not iron deficient, n = 116	5.28ˣ	37.0ˣ	12.0ˣ	—	70ˣ	23ˣ	32
Macaca mulatta[40]♂, 2–5 yr age, captive, bled 3x/wk, day 1, n = 20	5.2	43.0	13.8	—	83	27	32
Macaca mulatta[40]♂, 2–5 yr age, captive, bled 3x/wk, day 30, n = 20	2.90	23.0	6.1	—	79	21	26
Macaca mulatta[42], captive bred, 3.5 years old, indoor single housing, fasting 16 hr, touched and hand-to-hand fed daily, mean of 6 monthly analyses							
Macaca mulatta[42]♂, captive, adult, monkey chair, n = 6	5.89	43.0	13.8	—	73	23	32
Macaca mulatta[42]♀, captive, adult, monkey chair, n = 6	5.58	42.3	13.9	—	76	25	33

(Continued)

TABLE 1   193

Table 1 (Continued) Erythrocyte Counts and Related Values

	RBC	Hct	Hb	Diameter	MCV	MCH	MCHC
Macaca mulatta[43], captive, 3.5–5 years old, indoor single or paired housing, fasting 16–24 hr, studied over 4-year period, 2–15 analyses per subject							
Macaca mulatta[43] ♂, 3.5–6 kg, ketamine, n =53	5.32	38.9	12.6	—	73	24	32
Macaca mulatta[44] ♀, healthy, nonpregnant, adult 11–22 yr old, single housed							
Macaca mulatta[44] ♀, captive, n = 15, pre-ketamine	5.97×	44.7×	14.1×	—	75	24	31
Macaca mulatta[44] ♀, captive, n = 15, 15 min post-ketamine	5.50×	41.1×	13.0×	—	75	24	31
Macaca mulatta[45]♀, captive, adult, nonpregnant, squeeze-back cage, the same monkeys received each treatment 2 weeks apart							
Macaca mulatta[45]♀, restrained, no ketamine, zero time, n = 20	—	43.1	—	—	—	—	—
Macaca mulatta[45]♀, restrained, no ketamine, 10 min, n = 20	—	43.3	—	—	—	—	—
Macaca mulatta[45]♀, restrained, no ketamine, 20 min, n = 20	—	43.3×	—	—	—	—	—
Macaca mulatta[45]♀, restraint and ketamine, zero time, n = 20	—	42.0	—	—	—	—	—
Macaca mulatta[45]♀, restraint and ketamine, 10 min, n = 20	—	41.7	—	—	—	—	—
Macaca mulatta[45]♀, restraint and ketamine, 20 min, n = 20	—	40.1×	—	—	—	—	—
Macaca mulatta[46], captive, weight 2–3 kg, fasted, studied over 14 mo							
Macaca mulatta[46]♂, juvenile, n = 74	5.64	39.3	13.3	—	70	34	24
Macaca mulatta[46]♀, juvenile, n = 74	5.67	39.6	13.3	—	70	34	24
Macaca mulatta[47], newborn per C-section 10 days prior to predicted labor, mother administered glycopyrrolate, ketamine, and isoflurane gas							
Macaca mulatta[47]♂, umbilical cord blood, n = 10	4.18	39.9	13.1	—	94	31	33
Macaca mulatta[47]♀, umbilical cord blood, n = 7	4.35	41.8	13.6	—	96	31	33
Macaca mulatta[47]♂, and ♀combined, cord blood, n = 17	4.25	40.7	13.3	—	96	31	33
Macaca mulatta[48], captive >4 years old, analyzed per manual techniques							
Macaca mulatta[48] ♂, adult, n = 3	5.48	46.0	14.4	6	84	26	31
Macaca mulatta[48] ♀, adult, n = 4	5.33	40.9	13.3	6	78	25	32
Macaca mulatta[48], captive, >4 years old, analyzed per Coulter counter S							

(*Continued*)

Table 1 (Continued) Erythrocyte Counts and Related Values

	RBC	Hct	Hb	Diameter	MCV	MCH	MCHC
Macaca mulatta[48] ♂, adult, n = 3	6.37	47.2	15.1	6	74	23	32
Macaca mulatta[48] ♀, adult, n = 3	5.62	41.9	13.0	6	75	23	31
Macaca mulatta[50], laboratory-bred, hand-reared, neonatal longitudinal study							
Macaca mulatta[50], mixed sex, first 3 days of life, n = 168	—	49.1	—	—	—	—	—
Macaca mulatta[50], mixed sex, 1 week old, n = 116	—	40.9	—	—	—	—	—
Macaca mulatta[50], mixed sex, 2 weeks old, n = 107	—	37.2	—	—	—	—	—
Macaca mulatta[50], mixed sex, 3 weeks old, n = 118	—	37.3	—	—	—	—	—
Macaca mulatta[50], mixed sex, 4 weeks old, n = 118	—	37.2	—	—	—	—	—
Macaca mulatta[50], mixed sex, 8 weeks old, n = 118	—	37.8	—	—	—	—	—
Macaca mulatta[50], mixed sex, 12 weeks old, n = 117	—	39.0	—	—	—	—	—
Macaca mulatta[50], mixed sex, 16 weeks old, n = 118	—	38.9	—	—	—	—	—
Macaca mulatta[50], mixed sex, 20 weeks old, n = 117	—	39.2	—	—	—	—	—
Macaca mulatta[50], mixed sex, 24 weeks old, n = 122	—	39.7	—	—	—	—	—
Macaca mulatta[50], mixed sex, 28 weeks old, n = 121	—	40.3	—	—	—	—	—
Macaca mulatta[50], mixed sex, 32 weeks old, n = 118	—	40.7	—	—	—	—	—
Macaca mulatta[50], mixed sex, 36 weeks old, n = 124	—	41.1	—	—	—	—	—
Macaca mulatta[50], mixed sex, 40 weeks old, n = 121	—	41.1	—	—	—	—	—
Macaca mulatta[50], mixed sex, 44 weeks old, n = 115	—	41.2	—	—	—	—	—
Macaca mulatta[50], mixed sex, 48 weeks old, n = 116	—	41.0	—	—	—	—	—
Macaca mulatta[50], mixed sex, 52 weeks old, n = 115	—	41.4	—	—	—	—	—
Macaca mulatta[50], mixed sex, 18 months old, n = 105	—	41.3	—	—	—	—	—
Macaca mulatta[50], mixed sex, 2 years old, n = 37	—	42.7	—	—	—	—	—
Macaca mulatta[52], n = 6♂, 2♀, captive, adult, fasted, ketamine injections x3							
Macaca mulatta[52], day 1, post-ketamine	5.57	39.7	13.2	—	—	—	—

(Continued)

TABLE 1   195

Table 1 (Continued)   Erythrocyte Counts and Related Values

	RBC	Hct	Hb	Diameter	MCV	MCH	MCHC
Macaca mulatta[52], day 2, post-ketamine	5.46	38.6	12.8	—	—	—	—
Macaca mulatta[52], day 3, post-ketamine	5.38	37.9	12.5	—	—	—	—
Macaca mulatta[54]♀, 3–4 yr old, day 0, in cage, n = 10	4.78	36.5[ab]	11.6	—	76	24	32
Macaca mulatta[54]♀, 3–4 yr old, day 7, in chair, n = 10	5.04	38.3[ab]	11.8	—	76	23	31
Macaca mulatta[54]♀, 3–4 yr old, day 14, in chair, n = 10	4.82	36.1[b]	11.8	—	74	26	36
Macaca mulatta[54]♀, 3–4 yr old, day 21, in chair, n = 10	5.04	39.4[ab]	12.0	—	78	24	30
Macaca mulatta[54]♀, 3–4 yr old, day 28, in chair, n = 10	5.27	40.8[a]	12.3	—	78	23	30
Macaca mulatta[55] ♂, lab housed, 10–15 months old, n = 20	6.33	42.6	12.8	—	68	20	30
Macaca mulatta[55] ♀, lab housed, 10–15 months old, n = 13	6.07	42.5	12.9	—	70	22	30
Macaca mulatta[56] primate aging database, fasting, primarily indoor housed, some animals ketamine-injected, data controlled for sex							
Macaca mulatta[56], 7 years old	5.29	40.4	12.8	—	76	24	—
Macaca mulatta[56], mean for interval 6–30 years of age, n = 231–246	5.59	41.4	13.3	—	75	24	32
Macaca mulatta[57]♀, captive colony, 4–6 yr old, ketamine and xylazine							
Macaca mulatta[57]♀, between 6th and 10th week of pregnancy, n = 14	6.28	48.5	14.5	—	72	23	32
Macaca mulatta[57]♀, nonpregnant, n = 14	6.55	47.6	15.5	—	73	24	33
Macaca mulatta[58]♀, singly housed, maintained in a breeding colony							
Macaca mulatta[58]♀, mature, nonpregnant, controls n = 77	5.84	41.8	13.5	—	72	—	—
Macaca mulatta[58]♀, third trimester, n = 15–20	5.57	43.3	13.2	—	78	—	—
Macaca mulatta[58]♀, 24 hr postpartum, n = 18	4.85	37.3	11.5	—	77	—	—
Macaca mulatta[58]♀, 48 hr postpartum, n = 20	4.62	34.9	10.7	—	76	—	—
Macaca mulatta[58]♀, 72 hr postpartum, n = 19	4.57	34.4	10.4	—	75	—	—
Macaca mulatta[58]♀, 96 hr postpartum, n = 20	4.66	35.0	10.7	—	75	—	—
Macaca mulatta[58]♀, 1 week postpartum, n = 18	4.68	36.7	11.3	—	78	—	—
Macaca mulatta[58]♀, 2 weeks postpartum, n = 14	5.28	40.4	12.7	—	76	—	—

(Continued)

Table 1 (Continued) Erythrocyte Counts and Related Values

	RBC	Hct	Hb	Diameter	MCV	MCH	MCHC
Macaca mulatta[58]♀, 3 weeks postpartum, n = 20	5.55	41.8	13.1	—	75	—	—
Macaca mulatta[58]♀, 8 weeks postpartum, n = 20	5.78	42.4	13.4	—	73	—	—
Macaca mulatta[58]♀, singly housed, monitored during pregnancy and postpartum							
Macaca mulatta[58]♀, 17–53 days pregnant, n = 200	—	39.9	—	—	—	—	—
Macaca mulatta[58]♀, 54–104 days pregnant, n = 200	—	43.9	—	—	—	—	—
Macaca mulatta[58]♀, 105–131 days pregnant, n = 100	—	43.5	—	—	—	—	—
Macaca mulatta[58]♀, 132–140 days pregnant, n = 100	—	43.1	—	—	—	—	—
Macaca mulatta[58]♀, 141–163 days pregnant, n = 50	—	42.9	—	—	—	—	—
Macaca mulatta[58]♀, 1–12 hours postpartum, n = 175	—	37.9	—	—	—	—	—
Macaca mulatta[58]♀, 22–60 hours postpartum, n = 120	—	38.9	—	—	—	—	—
Macaca mulatta[58]♀, 20–28 days postpartum, n = 200	—	42.1	—	—	—	—	—
Macaca mulatta[58]♀, 46–77 days postpartum, n = 200	—	41.4	—	—	—	—	—
Macaca mulatta[77]♀, 29.6 mo old, control diet, at start, n = 8	4.99[a]	34.5[b]	11.8[c]	—	69[d]	24	—
Macaca mulatta[77]♀, 29.6 mo, control diet, at the end of 3 months, n = 8	4.47[a]	31.5[b]	10.3[c]	—	71[d]	23	—
Macaca mulatta[77]♀, 29.6 mo, Fe- & Zn-deficient diet, at start, n = 16	5.17[v]	35.4[w]	12.1[x]	—	69[y]	24[z]	—
Macaca mulatta[77]♀, 29.6 mo, deficient diet, at the end of 3 months, n = 16	4.65[v]	32.2[w]	10.6[x]	—	70[y]	23[z]	—
Macaca mulatta[78], 1♂, 2♀, 1.5 years old, iron-deficient diet	5.9	27–29	7	—	37	—	—
Macaca mulatta[78], 2♂, 1.5 years old, control, normal diet	6.0	46–49.	14	—	67	—	—
Macaca fascicularis, cynomolgus monkey, crab-eating or long-tailed macaque							
Macaca fascicularis[1], cynomolgus monkey, crab-eating or long-tailed macaque, captive colony, ketamine							
Macaca fascicularis[1] ♂, adult, 45–192 mo, mean 6,418 g, n = 21	6.30	39.8	13.6	—	64	22	34
Macaca fascicularis[1] ♀, adult, 45–192 mo, mean 4,490 g, n = 13	6.16	37.0	12.6	—	60	20	34

(Continued)

TABLE 1   197

Table 1 (Continued)   Erythrocyte Counts and Related Values

	RBC	Hct	Hb	Diameter	MCV	MCH	MCHC
Macaca fascicularis[8], cynomolgus monkey, crab-eating macaque, 5–6 yrs old, adult, before A.M. feeding							
Macaca fascicularis[8]♂, captive, mean 4.5 kg, adult, n = 8	6.86[x]	43.3	12.1[x]	—	63[x]	18[x]	28[x]
Macaca fascicularis[8]♀, captive, mean 3.0 kg, adult, n = 8	6.70[y]	41.6	11.7[y]	—	62[y]	17[y]	28[y]
Macaca fascicularis[9]♀, captive, >5 yrs old, mean 3.6 kg, fasting ≥16 hr, ketamine IM							
Macaca fascicularis[9]♀, captive, adult, n = 300	6.08	40.8	11.3	—	67	19	28
Macaca fascicularis[12]♂, mean 2.9 kg, captive bred, without ketamine, restrained							
Macaca fascicularis[12]♂, without ketamine, 3–5 yr old, n = 19	5.26[x]	47.5	12.7[x]	—	90[x]	24[x]	27[x]
Macaca fascicularis[12]♂, with ketamine, 3–5 yr old, n = 19	5.56[x]	47.0	13.6[x]	—	85[x]	24[x]	29[x]
Macaca fascicularis[12]♀, mean 2.7 kg, captive bred, without ketamine, restrained							
Macaca fascicularis[12]♀, without ketamine, 3–5 yr old, n = 15	5.13[x]	47.0[x]	12.2[x]	—	92[x]	24	26[x]
Macaca fascicularis[12]♀, with ketamine, 3–5 yr old, n = 15	5.35[x]	45.4[x]	12.9[x]	—	85[x]	24	28[x]
Macaca fascicularis[13], captive, 2–3.4 kg, fasted 16 hr, compared with *Macaca mulatta*[13]							
Macaca fascicularis[13]♂, captive. approx 4 yr old, n = 33	6.16[x]	41.0	11.8[x]	—	67	19	29
Macaca fascicularis[13]♀, captive, approx 4 yr old, n = 27	5.82[x]	39.0[x]	11.2[x]	—	67	19	29
Macaca fascicularis[14]♀, captive, after 6 mo lab environs, n = 222	6.50	42.0	11.7	—	65	18	28
Macaca fascicularis[17]♂, captive, 1 yr old	5.54	35.6	11.7	—	64	21	33
Macaca fascicularis[17]♀, captive, 1 yr old	5.76	37.3	11.9	—	65	21	32
Macaca fascicularis[17]♂, captive, 2 yr old	5.47	35.9	11.6	—	66	21	33
Macaca fascicularis[17]♀, captive, 2 yr old	6.03	39.5[x]	12.0	—	66	20	30[x]
Macaca fascicularis[17]♂, captive, 3–4 yr old	5.91	39.9[x]	13.0[x]	—	68[x]	22	33
Macaca fascicularis[17]♀, captive, 3–4 yr old	5.60	37.1	11.7	—	66	21	32
Macaca fascicularis[17]♂, captive, 5–6 yr old	5.62	37.1	12.2	—	66	22	33
Macaca fascicularis[17]♀, captive, 5–6 yr old	5.97	39.7[x]	12.1	—	67	20	31[x]

(Continued)

TABLE 1

Table 1 (Continued) Erythrocyte Counts and Related Values

	RBC	Hct	Hb	Diameter	MCV	MCH	MCHC
Macaca fascicularis[17]♂, captive, 7–8 yr old	5.56	37.2	12.4[x]	—	67[x]	23[x]	34
Macaca fascicularis[17]♀, captive, 7–8 yr old	5.76	38.5	12.1	—	67	21	32
Macaca fascicularis[17]♂, captive, 9–10 yr old	5.28	37.0	12.8[x]	—	71[x]	25[x]	35
Macaca fascicularis[17]♀, captive, 9–10 yr old	5.61	38.2	12.3	—	68[x]	22	32
Macaca fascicularis[17]♂, captive, 11–12 yr old	5.09[x]	37.9[x]	12.8[x]	—	75[x]	24[x]	34
Macaca fascicularis[17]♀, captive, 11–12 yr old	5.48	40.0[x]	13.0[x]	—	73[x]	25[x]	33
Macaca fascicularis[17]♂, captive, 13–14 yr old	5.28	39.3[x]	13.0[x]	—	74[x]	25[x]	33
Macaca fascicularis[17]♀, captive, 13–14 yr old	5.33	39.5	13.5[x]	—	74[x]	26[x]	34[x]
Macaca fascicularis[17]♂, captive, 15 yr old	5.26	40.5[x]	13.8[x]	—	77[x]	26[x]	34
Macaca fascicularis[17]♀, captive, 15 yr old	5.45	40.0[x]	13.3[x]	—	74[x]	25[x]	33[x]
Macaca fascicularis[17]♂, captive, mean 1–8 yr	—	—	—	—	66	—	—
Macaca fascicularis[17]♀, captive, mean 1–8 yr	—	—	—	—	66	—	—
Macaca fascicularis[17]♂, captive, mean 9–15 yr	—	—	—	—	74	—	—
Macaca fascicularis[17]♀, captive, mean 9–15 yr	—	—	—	—	72	—	—
Macaca fascicularis[18], fetal blood derived from repeat cardiocentesis of pregnant females immobilized with ketamine IM							
Macaca fascicularis[18], fetus gestation day 80, n = 7	3.02	33.0	11.3	—	109	37	34
Macaca fascicularis[18], fetus gestation day 90, n = 12	3.15	33.6	11.4	—	107	36	34
Macaca fascicularis[18], fetus gestation day 110, n = 14	3.61	36.3	12.4	—	101	35	34
Macaca fascicularis[18], fetus gestation day 120, n = 8	3.88	38.0	12.9	—	98	33	34
Macaca fascicularis[18], fetus gestation day 130, n = 10	4.11	38.2	12.8	—	94	33	33
Macaca fascicularis[18], fetus gestation day 140, n = 9	4.25	38.7	12.8	—	91	30	33
Macaca fascicularis[18], fetus gestation day 150, n = 8	4.03	36.7	12.4	—	91	31	34
Macaca fascicularis[20], captive, born, and bred at Tsukuba Primate Center							
Macaca fascicularis[20], captive, birth, n = 5	6.65	56.8	16.8	—	85	25	30

(Continued)

TABLE 1 199

Table 1 (Continued) Erythrocyte Counts and Related Values

	RBC	Hct	Hb	Diameter	MCV	MCH	MCHC
Macaca fascicularis[20], captive, age 1 day, n = 5	6.04	52.6	15.0	—	88	25	29
Macaca fascicularis[20], captive, age 2 days, n = 5	6.91	58.1	16.7	—	84	24	29
Macaca fascicularis[20], captive, age 3 days, n = 5	6.82	56.0	16.3	—	82	24	29
Macaca fascicularis[20], captive, age 4 days, n = 5	6.35	53.6	16.0	—	84	25	30
Macaca fascicularis[20], captive, age 5 days, n = 5	7.08	55.2	16.3	—	79	23	30
Macaca fascicularis[20], captive, age 6 days, n = 5	6.36	51.8	15.8	—	81	25	31
Macaca fascicularis[20], captive, age 7 days, n = 5	6.30	51.5	15.1	—	82	24	29
Macaca fascicularis[20], captive, age 1 month, n = 13	5.53	41.2	12.3	—	75	22	30
Macaca fascicularis[20], captive, age 2 months, n = 9	6.08	42.5	12.1	—	69	20	28
Macaca fascicularis[20], captive, age 3 months, n = 8	6.41	39.3	11.3	—	61	18	29
Macaca fascicularis[20], captive, age 4 months, n = 3	7.71	49.4	12.2	—	64	16	25
Macaca fascicularis[20], captive, age 5 months, n = 11	6.95	47.5	11.8	—	68	17	25
Macaca fascicularis[20], captive, age 6 months, n = 6	7.21	48.9	11.0	—	67	15	22
Macaca fascicularis[20], captive, age 7 months, n = 11	6.62	45.9	11.7	—	70	18	25
Macaca fascicularis[20], captive, age 8 months, n = 11	6.96	47.3	12.0	—	68	17	25
Macaca fascicularis[20], captive, age 9 months, n = 7	6.83	45.4	11.3	—	66	17	25
Macaca fascicularis[20], captive, age 10 months, n = 9	6.69	45.7	13.5	—	68	20	30
Macaca fascicularis[20], captive, age 11 months, n = 3	6.29	45.0	11.5	—	72	18	26
Macaca fascicularis[20], data of above							
Macaca fascicularis[20], mean 1st day post-birth thru 7 days	6.55	54.1	15.9	—	83	24	29
Macaca fascicularis[20], mean of months 2–6, n = 37	6.87	45.5	11.7	—	66	17	26
Macaca fascicularis[20], mean of months 7–11, n = 41	6.68	45.9	12.0	—	69	18	26
Macaca fascicularis[20], captive, C-section, cord blood, n = 6	4.98	44.1	12.7	—	89	26	29
Macaca fascicularis[20], femoral v, 5 hr after above, n = 6	5.91	49.9	14.9	—	84	25	30

(*Continued*)

Table 1 (Continued) Erythrocyte Counts and Related Values

	RBC	Hct	Hb	Diameter	MCV	MCH	MCHC
Macaca fascicularis[21] ♂ & ♀, captive, born, and bred at Tsukuba Primate Center, ketamine							
Macaca fascicularis[21] ♂, captive, 1 yr old, n = 20	6.51	44.3	12.3	—	68	19	28
Macaca fascicularis[21] ♀, captive, 1 yr old, n = 22	6.61	44.3	12.0	—	67	19	28
Macaca fascicularis[21] ♂, captive, 2 yr old, n = 20	6.60	41.8	11.8	—	63	18	28
Macaca fascicularis[21] ♀, captive, 2 yr old, n = 20	6.40	41.5	11.8	—	65	18	28
Macaca fascicularis[21] ♂, captive, 3–4 yr old, n = 20	5.99	43.7	12.3	—	73	21	28
Macaca fascicularis[21] ♀, captive, 3–4 yr old, n = 20	5.95	42.5	11.6	—	71	19	27
Macaca fascicularis[21] ♂, captive, 5–7 yr old, n = 20	6.03*	46.4*	13.2*	—	77	22	28
Macaca fascicularis[21] ♀, captive, 5–7 yr old, n = 23	5.60*	43.5*	12.3*	—	78	22	28
Macaca fascicularis[21] ♂, captive, 8–10 yr old, n = 2	6.71	52.4	15.0	—	78	22	29
Macaca fascicularis[21] ♀, captive, 8–10 yr old, n = 13	5.61	44.8	12.4	—	80	22	28
Macaca fascicularis[21] ♂, captive, 11–18 yr old, n = 9	6.08	50.1*	14.5	—	82	24	29
Macaca fascicularis[21] ♀, captive, 11–18 yr old, n = 17	5.72	44.1*	13.0	—	77	23	29
Macaca fascicularis[21] ♂, captive, mean of above, 1 through 4 years old	6.37	43.3	12.1	—	68	19	28
Macaca fascicularis[21] ♂, captive, mean of above, 1 through 4 years old	6.32	42.8	11.8	—	68	19	28
Macaca fascicularis[21] ♂, captive, mean of above, 5–18 yr old	6.27	49.6	14.2	—	79	22	29
Macaca fascicularis[21] ♀, captive, mean of above, 5–18 yr old	5.64	44.1	12.6	—	78	23	29
Macaca fascicularis[21] ♀, wild origin, ≥ 5 yr old, ketamine, n = 32	6.38	41.4	11.3	—	65	18	27
Macaca fascicularis[22] ♀, captive, bred, and reared at Tsukuba Primate Center, wt. > 3 kg, ketamine							
Macaca fascicularis[22] ♀, not pregnant, 6–34 years old, n = 78	5.91	40.3x	11.5x	—	68x	19	29
Macaca fascicularis[28], colony members, <1 to ≥5 yr old, ketamine IM, fasted 18 hr							
Macaca fascicularis[28] ♂, samplings: 36–45 (2 per subject)	6.00*	40.8	13.4	—	69*	23*	32
Macaca fascicularis[28] ♀, samplings: 46–49 (2 per subject)	5.50*	39.9	12.9	—	73*	24*	32
Macaca fascicularis[28], colony members, ketamine IM, fasted 18 hr							

(Continued)

TABLE 1   201

Table 1 (Continued) Erythrocyte Counts and Related Values

	RBC	Hct	Hb	Diameter	MCV	MCH	MCHC
Macaca fascicularis[28]♂, <1 yr old, samplings: 6–9 (2 per subject)	6.30	40.7	13.3	—	65^x	22	32
Macaca fascicularis[28]♂, 1–4 yr old, samplings: 18–23 (2 per subject)	5.90	40.0	13.1	—	68^y	22	32
Macaca fascicularis[28]♂, ≥5 yr old, samplings: 12–13 (2 per subject)	5.90	42.5	14.0	—	73^{xy}	24	33
Macaca fascicularis[28], colony members, ketamine IM, fasted 18 hr							
Macaca fascicularis[28]♀, <1 yr old, samplings: 7 (2 per subject)	6.40^{xy}	42.0^x	13.7^x	—	66^{xy}	22^x	33
Macaca fascicularis[28]♀, 1–4 yr old, samplings: 23–25 (2 per subject)	5.30^x	39.1^x	12.7^x	—	74^x	24^x	32
Macaca fascicularis[28]♀, ≥5 yr old, samplings: 16–17 (2 per subject)	5.30^y	40.1	12.8	—	74^y	24	32
Macaca fascicularis[28], colony members, <1 to ≥5 yr old, mixed sex, ketamine IM, fed normally							
Macaca fascicularis[28], samplings: 49–51 (once per subject)	5.90	39.9	12.2^x	—	68^x	21^x	30^x
Macaca fascicularis[28], colony members, <1 to ≥5 yr old, mixed sex, ketamine IM, fasted							
Macaca fascicularis[28], samplings: 82–94 (2 per subject)	5.70	40.3	13.1^x	—	71^x	23^x	32^x
Macaca fascicularis[37], plasma/serum	—	—	—	8.0	—	—	—
Macaca fascicularis[37], dry film smear	—	—	—	7.1	—	—	—
Macaca fascicularis[49], captive, 4–5 yr old, water avail *ad lib*							
Macaca fascicularis[49]♂, 3.1–6.0 kg, not fasting, n = 6	5.62	41.2	13.5	—	74	24	33
Macaca fascicularis[49]♀, 2.8–3.6 kg, not fasting, n = 5	5.02	36.0	11.8	—	72	24	33
Macaca fascicularis[49]♂, 3.1–6.0 kg, fasting 8 hr, n = 6	5.34	39.2	13.1	—	73	25	34
Macaca fascicularis[49]♀, 2.8–3.6 kg, fasting 8 hr, n = 5	5.08	36.4	12.0	—	72	24	33
Macaca fascicularis[49]♂, 3.1–6.0 kg, fasting 16 hr, n = 6	5.30^x	38.7^x	13.0^x	—	73^x	25^x	34^x
Macaca fascicularis[49]♀, 2.8–3.6 kg, fasting 16 hr, n = 5	4.94	35.3	11.8	—	72	24^x	34^x
Macaca fascicularis[49]♂, 3.1–6.0 kg, fasting 24 hr, n = 6	5.47	40.4	13.4	—	74	25	33
Macaca fascicularis[49]♀, 2.8–3.6 kg, fasting 24 hr, n = 5	5.07	36.7	12.1	—	72	24	33
Macaca fascicularis[53]♂, following stress of 15 hr transport from Japan to Korea							
Macaca fascicularis[53]♂, 3–4 yr, day 0, ketamine, n = 5	6.65^a	43.3^a	12.9^a	—	65^b	20^a	30^a

(Continued)

Table 1 (Continued) Erythrocyte Counts and Related Values

	RBC	Hct	Hb	Diameter	MCV	MCH	MCHC
Macaca fascicularis[53] ♂, 3–4 yr, day 7, ketamine, n = 5	5.82b	36.2b	11.3b	—	62b	19a	31a
Macaca fascicularis[53] ♂, 3–4 yr, day 14, ketamine, n = 5	5.66b	41.9a	11.1b	—	74a	20a	27b
Macaca fascicularis[53] ♂, 3–4 yr, day 21, ketamine, n = 5	5.37b	41.4a	10.6b	—	77a	20a	26b
Macaca fascicularis[53] ♂, 3–4 yr, day 28, ketamine, n = 5	5.51bc	42.6a	10.9b	—	77a	20a	26b
Macaca fascicularis[53] ♂, 3–4 yr, day 35, ketamine, n = 5	5.61b	41.4a	11.2b	—	74a	20a	27b
Macaca fascicularis[60], captive-bred, 2 yr old, Mauritius origin							
Macaca fascicularis[60] ♂, fasting, n = 15	7.21	50.2	14.3	—	70	20	29
Macaca fascicularis[60] ♀, fasting, n = 15	6.86	49.3	13.7	—	72	20	28
Macaca fascicularis[60], captured on Mauritius, est 2 yr old							
Macaca fascicularis[60] ♂, fasting, n = 15	6.89	49.3	13.9	—	72	20	28
Macaca fascicularis[60] ♀, fasting, n = 15	6.57	48.2	13.1	—	73	20	27
Macaca fascicularis[61], wild, Teluk Terima group of Bali Island							
Macaca fascicularis[61] ♂, ketamine, young adult and adult, n = 7	4.51	40.6	11.3	—	90	25	28
Macaca fascicularis[61] ♀, ketamine, young adult and adult, n = 8	4.15	39.3	11.0	—	95	27	28
Macaca fascicularis[61], wild, Pulaki group of Bali Island							
Macaca fascicularis[61] ♂, ketamine, young adult and adult, n = 10	5.16	46.9	13.5	—	91	26	29
Macaca fascicularis[61] ♀, ketamine, young adult and adult, n = 18	4.67	42.5	12.1	—	91	26	29
Macaca fascicularis[61], wild, Sangeh group of Bali Island							
Macaca fascicularis[61] ♂, ketamine, young adult and adult, n = 19	4.74	41.5	11.8	—	88	25	28
Macaca fascicularis[61] ♀, ketamine, young adult and adult, n = 11	4.86	40.6	11.7	—	84	24	29
Macaca fascicularis[61], wild, Ubud group of Bali Island							
Macaca fascicularis[61] ♂, ketamine, young adult and adult, n = 4	5.00	47.1	13.9	—	94	28	30
Macaca fascicularis[61] ♀, ketamine, young adult and adult, n = 4	4.87	45.2	13.3	—	93	27	30

(Continued)

TABLE 1 203

Table 1 (Continued) Erythrocyte Counts and Related Values

	RBC	Hct	Hb	Diameter	MCV	MCH	MCHC
Macaca fascicularis[61], wild, Kukuh group of Bali Island							
Macaca fascicularis[61] ♂, ketamine, young adult and adult, n = 1	4.11	40.3	11.9	—	98	29	30
Macaca fascicularis[61] ♀, ketamine, young adult and adult, n = 3	4.05	39.1	11.5	—	97	28	29
Macaca fascicularis[61], wild, Gunung Meru, West Sumatra							
Macaca fascicularis[61] ♂, ketamine, young adult and adult, n = 17	4.41	34.7	9.6	—	79	22	28
Macaca fascicularis[61] ♀, ketamine, young adult and adult, n = 23	4.30	33.2	9.0	—	77	21	27
Macaca fascicularis[62] housed at BioPrim in Baziege, France. Subjects obtained from breeders in Mauritius, Philippines, and Vietnam. *n* = 272 (183 ♂ = 67%, 89 ♀ = 33%). 60% subjects were 2–3 yr old, ketamine	6.13	41.7	12.2	—	68	20	29
Macaca fascicularis[63] ♂, mean 3.87 kg, n = 27	6.78*	41.9*	12.6*	—	62	19	30
Macaca fascicularis[63] ♀, mean 3.14 kg, n = 15	6.22*	38.8*	11.9*	—	63	19	31
Macaca fascicularis[64] ♂, adult, fasting, ketamine, n = 32	5.80x	35.3	11.7	—	62x	20x	33
Macaca fascicularis[65], juvenile, 2.5–3.5 yr old, 2.9–3.7 kg							
Macaca fascicularis[65] ♂, no anesthesia, n = 35	6.34*	49.7	14.5	—	79	23	29
Macaca fascicularis[65] ♀, no anesthesia, n = 37	6.11*	48.3	13.9	—	79	23	29
Macaca fascicularis[66], 3.5 kg							
Macaca fascicularis[66] ♀, no anesthesia, 0 min, pre-saline n = 6	6.18	40.2	12.3	—	65	20	31
Macaca fascicularis[66] ♀, 0 min, pre-ketamine, n = 6	6.21	40.8	12.4	—	66	20	30
Macaca fascicularis[66] ♀, 10 min, control, post-saline n = 6	6.38	40.1	12.3	—	63	19	31
Macaca fascicularis[66] ♀, 10 min, post-ketamine, n = 6	5.79	38.0	11.5	—	66	20	30
Macaca fascicularis[66] ♀, 30 min, control, post-saline n = 6	6.22	40.2	11.8	—	65	19	29
Macaca fascicularis[66] ♀, 30 min, post-ketamine, n = 6	5.64	36.8	11.2	—	65	20	30
Macaca fascicularis[66] ♀, 60 min, control, post-saline n = 6	6.07	39.3	11.4	—	65	19	29
Macaca fascicularis[66] ♀, 60 min, post-ketamine, n = 6	5.99	39.2	11.7	—	65	20	30

(*Continued*)

Table 1 (Continued) Erythrocyte Counts and Related Values

	RBC	Hct	Hb	Diameter	MCV	MCH	MCHC
Macaca fascicularis[66]♀, 120 min, control, post-saline n = 6	5.82	37.7	11.7	—	65	19	29
Macaca fascicularis[66]♀, 120 min, post-ketamine, n = 6	6.09	39.6	11.8	—	65	19	30
Macaca fascicularis[66]♀, 180 min, control, post-saline n = 6	5.46	35.5	11.0	—	65	20	31
Macaca fascicularis[66]♀, 180 min, post-ketamine, n = 6	6.15	39.8	11.9	—	65	19	30
Macaca fascicularis[67], fasting subjects, same sex, group housing indoors							
Macaca fascicularis[67]♂, 13–24 months age, no anesthesia, n = 162	5.78	44.9	13.0	—	78	22	29
Macaca fascicularis[67]♀, 13–24 months age, no anesthesia, n = 162	5.75	44.6	12.8	—	78	22	29
Macaca fascicularis[67]♂, 25–36 months age, no anesthesia, n = 121	5.86	45.5	13.1	—	78	22	29
Macaca fascicularis[67]♀, 25–36 months age, no anesthesia, n = 178	5.80	45.8	13.0	—	79	22	28
Macaca fascicularis[67]♂, 37–48 months age, no anesthesia, n = 16	5.87	45.5	13.0	—	78	22	29
Macaca fascicularis[67]♀, 37–48 months age, no anesthesia, n = 72	5.76	45.9	13.0	—	80	23	28
Macaca fascicularis[67]♂, 49–60 months age, no anesthesia, n = 31	5.89*	46.5*	13.2*	—	79	22	28
Macaca fascicularis[67]♀, 49–60 months age, no anesthesia, n = 84	5.55*	44.4*	12.5*	—	80	23	28
Macaca fascicularis[67]♂, 61–72 months age, no anesthesia, n = 44	5.90*	46.9*	13.3*	—	80	23	29
Macaca fascicularis[67]♀, 61–72 months age, no anesthesia, n = 47	5.32*	42.9*	12.1*	—	81	23	28
Macaca fascicularis[68]♂, 3–7 years old, no anesthesia, n = 95	5.49	42.9	13.2	—	78	24	31
Macaca fascicularis[68]♀, 3–7 years old, no anesthesia, n = 95	5.44	42.2	12.9	—	78	24	31
Macaca fascicularis[69], bred and reared in captivity, individually caged indoors							
Macaca fascicularis[69]♂, 12–14 months old, n = 42	6.57	44.0	12.0	—	67	—	—
Macaca fascicularis[69]♀, 12–14 months old, n = 34	6.65	44.5	12.0	—	68	—	—
Macaca fascicularis[69]♂, 15–17 months old, n = 60	6.44	42.9	11.7	—	66	—	—
Macaca fascicularis[69]♀, 15–17 months old, n = 60	6.50	43.3	11.8	—	66	—	—
Macaca fascicularis[69]♂, 18–20 months old, n = 60	6.46	42.3	11.8	—	65	—	—
Macaca fascicularis[69]♀, 18–20 months old, n = 60	6.39	42.8	11.8	—	67	—	—

(Continued)

Table 1 (Continued) Erythrocyte Counts and Related Values

	RBC	Hct	Hb	Diameter	MCV	MCH	MCHC
Macaca fascicularis[69]♂, 21–23 months old, n = 41	6.48	43.2	11.9	—	65	—	—
Macaca fascicularis[69]♀, 21–23 months old, n = 40	6.44	42.9	11.6	—	66	—	—
Macaca fascicularis[69]♂, 24–26 months old, n = 57	6.51	43.1	11.7	—	66	—	—
Macaca fascicularis[69]♀, 24–26 months old, n = 48	6.42	43.5	11.8	—	68	—	—
Macaca fascicularis[69]♂, 27–29 months old, n = 46	6.44	43.7	11.9	—	68	—	—
Macaca fascicularis[69]♀, 27–29 months old, n = 57	6.55	43.8	11.7	—	68	—	—
Macaca fascicularis[69]♂, 30–32 months old, n = 57	6.72	46.2	12.3	—	68	—	—
Macaca fascicularis[69]♀, 30–32 months old, n = 57	6.37	44.8	11.9	—	70	—	—
Macaca fascicularis[69]♂, 33–35 months old, n = 22	6.48	44.7	12.1	—	69	—	—
Macaca fascicularis[69]♀, 33–35 months old, n = 25	6.47	43.7	11.9	—	67	—	—
Macaca fascicularis[69]♂, 36–38 months old, n = 11	6.44	44.4	12.3	—	69	—	—
Macaca fascicularis[69]♀, 36–38 months old, n = 12	6.08	41.8	11.2	—	69	—	—
Macaca fascicularis[69]♂, 39–41 months old, n = 19	6.37	43.5	11.8	—	68	—	—
Macaca fascicularis[69]♀, 39–41 months old, n = 27	6.49	44.1	12.2	—	67	—	—
Macaca fascicularis[69]♂, 42–44 months old, n = 15	6.47	45.2	12.5	—	70	—	—
Macaca fascicularis[69]♀, 42–44 months old, n = 19	6.51	47.5	12.5	—	73	—	—
Macaca fascicularis[69]♂, 45–47 months old, n = 8	6.99	51.6	13.9	—	73	—	—
Macaca fascicularis[69]♀, 45–47 months old, n = 12	5.91	44.3	11.9	—	74	—	—
Macaca fascicularis[69]♂, 48–53 months old, n = 13	6.66	48.2	13.2	—	74	—	—
Macaca fascicularis[69]♀, 48–53 months old, n = 16	6.38	46.7	12.8	—	73	—	—
Macaca fascicularis[69]♂, 54–59 months old, n = 11	6.15	45.8	12.6	—	75	—	—
Macaca fascicularis[69]♀, 54–59 months old, n = 9	5.87	44.9	12.3	—	76	—	—
Macaca fascicularis[69]♂, 60–71 months old, n = 21	5.84	45.7	12.5	—	79	—	—
Macaca fascicularis[69]♀, 60–71 months old, n = 25	5.69	43.8	12.2	—	77	—	—
Macaca fascicularis[69]♂, 72–83 months old, n = 19	5.97	45.6	12.9	—	76	—	—

(Continued)

Table 1 (Continued) Erythrocyte Counts and Related Values

	RBC	Hct	Hb	Diameter	MCV	MCH	MCHC
Macaca fascicularis[69]♀, 72–83 months old, n = 11	5.68	43.9	12.6	—	77	—	—
Macaca fascicularis[69]♂, 84–95 months old, n = 10	6.09	45.2	13.1	—	74	—	—
Macaca fascicularis[69]♀, 84–95 months old, n = 9	5.24	42.8	11.8	—	81	—	—
Macaca fascicularis[69]♂, 96–108 months old, n = 7	6.26	47.7	14.0	—	75	—	—
Macaca fascicularis[69]♀, 96–108 months old, n = 14	5.55	43.2	12.2	—	78	—	—
Macaca fascicularis[69]♂, >109 months old, n = 12	5.74	47.2	13.6	—	82	—	—
Macaca fascicularis[69]♀, >109 months old, n = 20	5.69	44.6	12.6	—	78	—	—
Macaca fascicularis[70], control normal, at 5th mo study, n = 3	6.20	41.0	11.4	—	63	19	31
Macaca fascicularis[70], vitamin E–deficient, at 5th mo study, n = 3	6.40	36.7	10.7	—	63	20	31
Macaca fascicularis[70], control normal, at 18th mo study, n = 3	6.30[a]	41.7[a]	11.1[a]	—	64	19	31
Macaca fascicularis[70], vitamin E–deficient, at 18th mo study, n = 3	4.00[a]	25.0[a]	6.7[a]	—	63	20	31
Macaca fascicularis[71]♂, Charles River Laboratories	5.90	35.9	11.7	—	61	20	32
Macaca fascicularis[71]♀	5.40	32.0	10.2	—	60	19	32
Macaca fascicularis[72]♀, control, 23 nongravid, day 11, menstrual cycle	5.20	38.7	13.3	—	—	—	—
Macaca fascicularis[72]♀, 0 wk of pregnancy, n = 14	5.14	38.9	13.3	—	—	—	—
Macaca fascicularis[72]♀, 2 wk of pregnancy, n = 14	5.05	39.2	13.5	—	—	—	—
Macaca fascicularis[72]♀, 4 wk of pregnancy, n = 14	5.07	38.1	13.2	—	—	—	—
Macaca fascicularis[72]♀, 6 wk of pregnancy, n = 14	5.13	38.9	13.5	—	—	—	—
Macaca fascicularis[72]♀, 8 wk of pregnancy, n = 14	5.28	38.9	13.9	—	—	—	—
Macaca fascicularis[72]♀, 10 wk of pregnancy, n = 14	4.95	39.3	13.4	—	—	—	—
Macaca fascicularis[72]♀, 12 wk of pregnancy, n = 14	5.12	39.8	13.6	—	—	—	—
Macaca fascicularis[72]♀, 14 wk of pregnancy, n = 14	5.21	38.6	13.9	—	—	—	—
Macaca fascicularis[72]♀, 16 wk of pregnancy, n = 14	4.97	38.4	13.5	—	—	—	—
Macaca fascicularis[72]♀, 4 wk prepartum, n = 14	5.05	38.3	13.2	—	—	—	—

(Continued)

TABLE 1 207

Table 1 (Continued) Erythrocyte Counts and Related Values

	RBC	Hct	Hb	Diameter	MCV	MCH	MCHC
Macaca fascicularis[72]♀, 2 wk prepartum, n = 14	4.60	36.9	12.6	—	—	—	—
Macaca fascicularis[72]♀, 1 wk prepartum, n = 14	4.80	35.9	12.6	—	—	—	—
Macaca fascicularis[72]♀, 1 wk postpartum, n = 14	4.21	35.0	11.3	—	—	—	—
Macaca fascicularis[72]♀, 2 wk postpartum, n = 14	4.50	36.4	11.8	—	—	—	—
Macaca fascicularis[72]♀, 4 wk postpartum, n = 14	4.21	40.4	13.0	—	—	—	—
Macaca fascicularis[72]♀, 8 wk postpartum, n = 14	4.21	42.9	13.7	—	—	—	—
Macaca fascicularis[73], adult, laboratory housed, studied over 3 yr							
Macaca fascicularis[73]♂, ketamine, n = 89	—	40.9	12.3	—	—	—	30
Macaca fascicularis[73]♀, ketamine, n = 53	—	39.4	12.0	—	—	—	29
Macaca fascicularis[74]♂, all ages, unanesthetized, n = 386	6.64	40.6	12.3	—	61	19	30
Macaca fascicularis[74]♀, all ages, unanesthetized, n = 397	6.29	37.9	11.4	—	60	18	30
Macaca fascicularis[75]♂, 2 mo to 15 yr, ketamine, in colony, n = 35	6.56*	38.0*	12.0*	—	59	18	31
Macaca fascicularis[75]♀, 1.5 mo to 8 yr, ketamine, in colony, n = 31	6.00*	36.0*	11.0*	—	60	19	31
Macaca fascicularis[76]♂, 2–5 yr, multiple institutions, n = ~540	6.48	41.5	12.1	—	—	—	—
Macaca fascicularis[76]♀, 2–5 yr, multiple institutions, n = 500	6.19	40.1	11.5	—	—	—	—

Notes: RBC, mean erythrocyte count, millions/mm³ or μL; Hct, mean hematocrit; Hb, mean grams hemoglobin/deciliter; Diameter, mean diameter (μm) in dry film smears unless otherwise indicated; MCV, mean cellular volume (μm³ or femtoliters, fL); MCH, mean cellular hemoglobin (pg); MCHC, mean cellular hemoglobin concentration (wt./vol%, g Hb/100 mL RBC).

Asterisk (*) indicates a significant difference between sexes for a given parameter. Unless otherwise indicated, superscript letters identify significant differences between values for identified parameters derived for compared groups of subjects. Some erythrocyte indices have been calculated from reported data per Wintrobe formulas.

(Continued)

Table 1 (*Continued*) Erythrocyte Counts and Related Values

	RBC	Hct	Hb	Diameter	MCV	MCH	MCHC

1 Andrade et al. (2004) (Long-term captive clinically healthy primates at Centro de Criacao de Animais de Laboratorio, Departamento de Primatologia-Fiocruz, Rio de Janeiro, RJ, Brazil. They were maintained in harem groups of 1 male for 8–10 females. *Macaca mulatta* and *Macaca fascicularis* were housed outdoors while *Saimiri sciureus* were maintained in a building receiving natural light. They were anesthetized with ketamine hydrochloride and blood was obtained via femoral puncture. The animals were phlebotomized once a year. The monkeys were divided into five age groups based on their development and reproductive capacity: [1] baby monkeys, 0–6 months; [2] infants, post-weaned, 7–18 months, [3] juveniles, 19–31 months, [4] young monkeys, puberty, 32–44 months, [5] sexually mature, 45–192 months. RBC counts and hemoglobin concentrations were determined electronically [530/550 CELM, Cia Equipadora de Laboratorios Modernos] while the hematocrits were established by centrifugation.)

2 King et al. (1967), also listed in Morrow et al. (1970) (Normal mature female subjects. The monkeys were obtained from a commercial importer and were isolated for 3 months prior to the study. They were housed in hanging-type cages. Determinations were performed per femoral phlebotomy at the initiation of the study and at 45 and 90 days, and thereafter at 90-day intervals until sacrifice. The interval was extended to 120 days during the second and third years. The observation periods ranged from 3 to 33 months with a mean of 19 months. The RBC counts were performed with a Coulter counter, Hb was quantitated spectrophotometrically as cyanmethemoglobin, and the Hct was determined per the microhematocrit technique. The erythrocyte indices were derived from the reported data with the application of the Wintrobe formulas.)

3 Robinson et al. (1968), also listed in Morrow et al. (1970) (Normal young adults. The subjects were placed on a 90-day stabilization period after their purchase from the commercial supplier. Several purchases were made. The weights of the 42 males ranged from 2.7 to 9.1 kg [mean 4.3 kg]. The weights of the females ranged from 2.3 to 6.8 kg (mean 3.7 kg). In 82% of the subjects three or four 8–10 mL femoral venous samplings were taken over the 6-month study period. The RBC counts were performed with a Coulter counter, Hb was quantitated spectrophotometrically as cyanmethemo-globin, and the Hct was determined per the microhematocrit technique.)

4 Switzer (1967a), also listed in Morrow et al. (1970) (The subjects were clinically normal adults; male monkeys weighed 5.4–7.6 kg and had been in the colony for 9 months to 1 year. Females weighed 3.7–8.2 kg and had been in the colony for 1–4 years. The subjects were individually caged. The mean age of the males was 5.8 yr and that of the females was 8.2 yr. The females were at least 60 days postpartum. Blood was obtained by femoral veni-puncture. The RBC counts were performed with a Coulter counter, Hb was quantitated spectrophotometrically as cyanmethemoglobin, and the Hct was determined per the microhematocrit technique.)

(Continued)

TABLE 1 209

Table 1 (*Continued*) Erythrocyte Counts and Related Values

	RBC	Hct	Hb	Diameter	MCV	MCH	MCHC

5   Stanley et al. (1968) (The monkeys were wild-caught in the highlands of northern India. They were given a 72-day conditioning period before hematologic assessments. Subjects were maintained individually in primate cages. One group [5a] consisted of five male and six female formerly wild Indian subjects that were 3.5–6 years of age and weighed 4.5–6 kg. They were sampled weekly for 40 weeks per 5 mL femoral artery phlebotomies [~1.7% estimated total body blood volume per sample]. A second group [5b] and [5c] consisted of 74 male and 56 female similarly derived and maintained captive monkeys. They were assigned to groups for each sex and age. Distribution: males less than 3 years of age, 24–32 mo of age, n = 36; females less than 3 years of age, 24–32 mo of age, n = 22 (group [5b]); males over 3 years of age, 36–60 mo, n = 38; and females over 3 years of age, 36–60 mo, n = 34 (group [5c]). The members of this group ([5b] and [5c]) were phlebotomized per femoral vein every other week for a period of 12 weeks. Blood samples were collected for all groups before the morning feeding. The erythrocyte counts were determined with an electronic counter; the hemoglobin concentration was assayed spectrophotometrically as cyanmethemoglobin; and the hematocrit was derived by the microhematocrit technique. The superscript "x" indicates a significant difference in values for indicated parameters for males less than 3 years of age versus males older than 3 years of age.)

8   Matsumoto et al. (1980) (*Macaca mulatta* were obtained from India and were an average estimated 7–8 years old. Their average weight for males was 7.0 kg and the mean weight for females was 4.9 kg. *Macaca fascicularis* were obtained from Malaysia and were an estimated 5–6 years old; the males weighed an average 4.5 kg while the weight for the females was a mean 3.0 kg. The hemograms including the Hct were determined with a Coulter counter. Blood samples were obtained before the A.M. feeding from the saphenous vein. The animals were phlebotomized while held in a specially designed restraining device. Ketamine or other anesthetic drugs were apparently not used. The asterisk indicates a significant difference between the values for the indicated parameter in the sexes of *Macaca mulatta*. The superscript "x" indicates a significant difference between the values for the indicated parameters between the males of *M. mulatta* and *M. fascicularis*. The superscript "y" indicates a significant difference between the values for the indicated parameters between the females of *M. mulatta* and *M. fascicularis*.)

9   Yoshida et al. (1989) (Subjects were obtained from Indonesia, Philippines, and Malaysia and had been maintained in individual cages under uniform environmental conditions for 4–5 years at the Tsukuba Primate Center for Medical Science, National Institute of Health, Japan. The monkeys were an estimated age of more than 5 years of age. The animals were sedated with ketamine hydrochloride 5 mg/kg, and the blood samples were obtained by femoral puncture. The erythrocyte counts, hemoglobin concentrations [as cyanmethemoglobin], hematocrits, and mean red cell volumes were determined by microcell counter CC-108, 110, MH-301, Toa Medical Electronics Co. The MCH and MCHC were derived from the reported data with the application of the Wintrobe formulas.)

(*Continued*)

Table 1 (*Continued*) Erythrocyte Counts and Related Values

	RBC	Hct	Hb	Diameter	MCV	MCH	MCHC

[12] Kim et al. (2005b) (Analyses were conducted on 19 male and 15 female 3–5 years of age captive-bred subjects derived from Guangxi Primate Center, China. The mean weight of the males was 2.9 kg and that of females was 2.7 kg. The analyses were conducted after intramuscular administration of ketamine 10 mg/kg, which was administered while the monkeys were in their individual cages. Blood samples were obtained from the cephalic vein within 30 minutes after the infection of the drug. The same monkeys were sampled again per cephalic vein 19 days later but in this instance they were not anesthetized and were phlebotomized while seated in a monkey chair after physical restraint with the squeezer in an individual's cage. The analyses were conducted electronically with a hematologic autoanalyzer, ADVIA 120 Hematology System. The Hct was determined from the MCV. The superscript "x" indicates a significant difference in values for the cited parameters for the indicated sex in monkeys given ketamine prior to phlebotomy versus the same group of subjects not administered the anesthetic prior to the sampling of the blood.)

[13] Matsuzawa et al. (1994) (A comparison of the erythrograms of purpose-bred/colony-bred monkeys, *Macaca fascicularis* cynomolgus monkeys, and *Macaca mulatta* rhesus monkeys. The subjects were housed under identical conditions for 9 months in the primate colony, Safety Research Laboratories, Yamanouchi Pharmaceutical Company, Tokyo, Japan. Cynomolgus monkeys were housed individually in stainless steel cages; the animals were approximately 4 years old and weighed 2–3.4 kg. The rhesus monkeys were also housed individually in stainless cages; the animals were approximately 3 years old and weighed between 2.2 and 3.5 kg. Blood samples were obtained from the femoral or saphenous veins while the subjects were conscious; the subjects were starved 16 hr prior to phlebotomy. The hematocrits were determined by microhematocrit centrifugation, while the RBC counts and hemoglobin concentrations of the blood were measured electronically with a Coulter counter. The erythrocyte indices were derived from the reported data with the application of the Wintrobe formulas. Reticulocytes were quantitated microscopically by the Brecher method. The RBC count, Hct, and Hb concentration of the blood of the cynomolgus monkeys were statistically compared with those of the rhesus monkeys. Statistical differences between cynomolgus versus rhesus monkeys are indicated by the superscript "x.")

[14] Yoshida (1981) (Analyses conducted 6 months after admission into Tsukuba Primate Center for Medical Science, National Institute of Health, Japan. Blood samples were obtained per femoral phlebotomy. RBC, Hct, and Hb were determined electronically [Toa Microcell Counter System]. The erythrocyte indices were derived from the reported data with the application of the Wintrobe formulas.)

(Continued)

TABLE 1

211

Table 1 (*Continued*) Erythrocyte Counts and Related Values

0.5 pt	RBC	Hct	Hb	Diameter	MCV	MCH	MCHC

[16] Rosenblum et al. (1981) [Two populations of rhesus monkeys, 32 sexually mature adults, and 48 sexually immature juveniles [approximately 2 years old on arrival at the facility] were subjected to hemogramic analyses. In addition, a longitudinal 1-year long study was conducted on five male and five female juveniles. Analyses were performed every 3 months. A 3-year long survey of four adult sexually mature subjects of each sex was also conducted. Analyses were performed every 6 months. The monkeys were housed individually. The mean weights of the subjects are included in the tables. All animals were trained to present an arm through a small opening of the cage. Thus, the blood samples were drawn from untranquilized fully alert subjects. Blood was withdrawn in the morning after they had been fasted for at least 18 hr. The superscript "x" indicates a significant difference between the value for the indicated parameter derived at baseline and the value derived at the 12th month of study. Superscript "y" indicates a significant difference between the value for the indicated parameter derived at the 12th month of study and the preceding time interval. The superscript "a" indicates a significant difference between the values for the indicated parameter for the juvenile males versus that of adult males. The superscript "b" indicates a significant difference between the value for the indicated parameter derived at baseline and the value derived at the indicated month of study. The RBC counts were performed with a Coulter counter, Hb was quantitated spectrophotometrically as cyanmethemoglobin, and the hematocrits were determined per centrifugation. The erythrocyte indices were derived from the reported data with the application of the Wintrobe formulas.)

[17] Terao (2005) (The subjects were laboratory bred and maintained at the Tsukuba Primate Center for Medical Science, National Institute of Infectious Diseases, Japan. The superscript "x" indicates a significant difference between the value for the indicated parameter derived at 1 year and the value derived at the indicated age.)

[18] Tarantal (1993) (Hemograms were conducted on fetal *Macaca fascicularis* of established fetal ages by repeat ultrasound-guided cardiocentesis. Studies were conducted at the California Regional Primate Research Center, University of California at Davis. Thirty-one fetuses were sampled for a total of 68 samplings, 2–5 phlebotomies/fetus, during gestational days 80–150. The pregnant females were immobilized with ketamine, 10 mg/kg IM. Hemograms were established electronically with a Serono Baker Diagnostic System, Allentown, PA.)

[19] Woodward et al. (1997) (Nine male and nine female captive 3-year-old individually housed *Macaca mulatta* weighing 3–4.5 kg were anesthetized and phlebotomized per femoral vein on three occasions at 10-day intervals. The monkeys were randomly divided into three groups of six macaques [three males and three females/group]. Each group was anesthetized with each different immobilization agent for the series of three blood samplings. The agents were ketamine hydrochloride 10 mg/kg IM, ketamine 10 mg/kg, and acepromazine maleate 2 mg/kg IM, or telazol 3 mg/kg IM [a combination of tiletamine and zolazepam]. The hemograms were performed electronically using a Coulter STK-S instrument.)

[20] Sugimoto et al. (1986b) (Newborn, infant, and older cynomolgus monkeys born and reared at Tsukuba Primate Center for Medical Science, Japan. The RBC count, Hct, Hb concentration of the blood and MCV were determined with electronic instrumentation, Toa Medical Electronics type CC-108, and CC-110. The MCH and MCHC were derived from the reported data with the application of the Wintrobe formulas. Blood was obtained from the femoral vein without anesthesia. Cord blood was obtained from six additional animals immediately after cesarean section followed by a femoral vein aspiration 5 hours later.)

(Continued)

Table 1 (Continued) Erythrocyte Counts and Related Values

	RBC	Hct	Hb	Diameter	MCV	MCH	MCHC

[21] Sugimoto et al. (1986a) (206 monkeys bred and reared at Tsukuba Primate Center for Medical Science, National Institute of Health, Japan as well as 32 female cynomolgus monkeys of wild origin estimated to be 5 years of age or older were studied. Blood was withdrawn from the femoral vein under anesthesia with ketalar 5–10 mg/kg. The RBC count, Hct, and Hb concentration of the blood were determined with electronic instrumentation Toa Medical Electronics type CC-108 and CC-110. The erythrocyte indices were derived from the reported data with the application of the Wintrobe formulas.)

[22] Chen et al. (2002) (Cynomolgus monkeys bred and reared at Tsukuba Primate Center, National Institute of Infectious Diseases, Japan. The monkeys were nonpregnant females and ranged from 6 to 34 years of age. Their body weight was greater than 3 kg, mean 4.5 kg. The animals were anesthetized with 10 mg ketamine/kg. The RBC count, Hct, Hb concentration of the blood, and MCV were determined with electronic instrumentation, Toa Medical Electronics Sysmex F-800. MCH and MCHC were derived from the reported data with the application of the Wintrobe formulas. Superscript "x" indicates the parameters that were observed to be significantly correlated with body weight. Blood was obtained from the femoral vein.)

[23] Ives et al. (1956) (18 young male [mean initial weight 2.9 kg] and 18 young female [mean initial weight 2.8 kg] monkeys having been obtained from a conditioning farm in South Carolina were "familiarized" with the testing laboratory for 4 months prior to the initiation of the investigation which was conducted over a period of 12 months. The mean weight of the males at the end of the study was 5.1 kg, and the respective weight for the females was 4.6 kg. The exact ages of the monkeys were unknown. However it was estimated, on the basis of similar weights of both sexes at the onset of the study and a statistical difference between their weights at the end of the 12-month study, presumed to be a result of the appearance of a sexual difference in weight, that the monkeys were over 4 years old at the end of the test period. Blood was removed from the ear for the hemograms; hemoglobin was determined as cyanmethemoglobin while the method for establishing the erythrocyte counts was "that which was currently used in clinical laboratories." The monkeys were handled daily in the feeding of food supplements per gastric tube. MCH was derived from the reported data with the application of the Wintrobe formula.)

[24] Lilly et al. (1999) (45 female subjects were captured at 22–25 months of age [2.1–4.8 kg body weight, mean 3.6 kg], from a population of approximately 4500 individuals living in a free-ranging naturalistic environment. They inhabited a densely wooded barrier island and were food provisioned. The subjects were born and resided there until the onset of the investigation. Phase I of the study: subjects were randomly selected from individuals trapped in capture corrals over a period of 5 days. They had access to their normal diet during and before capture. The selected corral-captured subjects were caught using a capture net and were anesthetized with an average of 70 mg of ketamine. Blood samples were generally obtained within 50 min after capture. Phase I, the day of capture and phlebotomy was considered a period of acute stress. Phase II: the subjects were transported to a mainland facility and housed singly in indoor cages in single caging. They were socially isolated from their peers and had no tactile contact but had visual and auditory contact with some members of the study. They were caught using squeeze-back cages and were phlebotomized at five regular intervals in a 6-month period. Ketamine was administered and blood sampling occurred about 22 min later. Phase II was considered a period of chronic stress. Phase III: each monkey was housed with four other females. This interval was considered as one offering a normalized environment. The first sample was taken after the groups became socially stable, about 3 months after the individuals were placed in the groups. The second sample was obtained 3 months later. Blood samples were collected between 40 and 45 minutes after entry of the investigators into the housing area and the subjects were caught by net. Ketamine was presumed to have been administered. Complete blood counts were performed with the electronic instrument Technicon H-1.)

(Continued)

TABLE 1　　　　213

Table 1 (Continued)　Erythrocyte Counts and Related Values

	RBC	Hct	Hb	Diameter	MCV	MCH	MCHC

25　Kessler et al. (1983) (24-year-old rhesus monkeys were analyzed. Seven were males 15–28 years old [mean 20.3 yr] and 17 were females 16–25 years old [mean 19.6 yr]. The mean weight of the males was 8.9 kg and that of the females was 7.9 kg. They had been maintained as members of a free-ranging colony on the Island of La Cueva, Puertco Rico and transferred to outdoor enclosures in Sabana Seca in 1981. Values were also established for 34 younger adult monkeys maintained under identical conditions. These consisted of 10 adult males 6–14 years old [mean 7.5 yr, mean weight 9.4 kg], and 24 adult females 4–10 years old [mean 7.4 yr, mean weight 7.2 kg]. All monkeys were fasted overnight with access to water prior to obtaining the blood samples. The subjects were administered 10 mg ketamine/kg intramuscularly prior to femoral phlebotomy. Superscript "x" indicates a statistical difference for the indicated parameter between aged and adult monkeys of the same sex. RBC counts, Hb, Hct values as well as the erythrocyte indices were electronically determined on a Coulter counter model S.)

28　Perretta et al. (1991) (The investigation was conducted on a closed colony of monkeys that was established ~10 years prior to the investigation. The animals were maintained in breeding groups of 15–20 monkeys in indoor–outdoor facilities. The colony consisted of seven feral born individuals obtained from commercial suppliers and 49 colony-born individuals. The ages ranged from less than 1 year to more than 5 years of age. Three separate femoral vein samplings were obtained from each subject. For two of the samples, the monkeys were fasted for 18 hours prior to phlebotomy [food was withheld but water was presumably available ad lib.], while another sample was obtained from the monkeys when they were on the normal feeding schedule. Ketamine, 10 mg/kg, was administered intramuscularly prior to each sampling. The erythrocyte counts were derived electronically with a Coulter counter, the Hb was determined spectrophotometrically as cyanmethemoglobin while the Hct was established by microhematocrit centrifugation. The derived data were allocated to different subsets to compare the values obtained for males versus females, for comparison of different age categories of each sex, and to evaluate the data derived from pregnant females [≥5 years of age] with comparably aged, nonpregnant females. Data of fasted and fed subjects were also compared. The values of pregnant females were solely used in the analysis of pregnancy, and the data of fed subjects were used only in comparison with fasted monkeys. Overall evaluations of male versus female data or that of different age categories utilized values of fasting subjects only. Values for an identified parameter within a table for a given subgroup indicated by the superscript "x" [e.g., RBC of females <1-year old versus the RBC of females1–4 years old] are significantly different. The same statistical relationship is indicated for parameters of a given subgroup indicated by the superscript "y".)

29　Kupper et al. (1976) (The sampled population consisted of 56 male and 57 female healthy captive rhesus monkeys purchased as a group from another USA military research laboratory. The ages of the subjects were established by examination of the dentition. The males ranged in age from 24 to 54 months [mean 42 mo] while the females ranged from 24 to 80 months [mean 50 mo]. Males weighed 3.4–8.5 kg [mean 5.0 kg], females weighed 3.1–7.0 kg [mean 4.7 kg]. The subjects were deprived of food for 17 hr prior to femoral venous phlebotomy, but were allowed water ad lib. Thirty minutes prior to phlebotomy the monkeys were administered the neuroleptic–narcotic drug, fentanyl-droperidol [Innovar] 1 cc/18 kg, IM. The monkeys were restrained in a horizontal position. The blood was anticoagulated with heparin. The RBC count and Hct were determined using the model ZBI Coulter counter while the Hgb concentration was determined as cyanmethemoglobin using a Coulter hemoglobinometer.)

(Continued)

Table 1 (*Continued*) Erythrocyte Counts and Related Values

	RBC	Hct	Hb	Diameter	MCV	MCH	MCHC

30 Kessler and Rawlins (1983) (64 male and 33 female free-ranging rhesus monkeys from one of six social groups on the island of Cayo Santiago, Puerto Rico were studied. They were provisioned daily with commercial monkey diet which was supplemented by the subjects' independent foraging for seeds, berries, and so on, that naturally occurred in the subtropical ecosystem. The data are reported by sex for each of three age groups: 2–3, 4–9, and ≥10 years. All adult females (≥4 years) were pregnant. These monkeys represent one of the longest continually maintained colonies of free-ranging rhesus monkeys in captivity. The monkeys were isolated since their introduction from India in 1938. All of the monkeys were of known age, sex, and maternal genealogy. Yearling monkeys are trapped annually to be tattooed and ear notched for identification. In January 1981, one entire social group was captured during the regular trapping period. This was the studied group. The yearlings were not bled for this study because of prior research commitments. The subjects were weighed and examined, and all were in excellent health. The birth season on Cayo Santiago extends from late December until early July with a modal birth date in March. As a consequence, the majority of the females were in midterm pregnancy. Each monkey was captured in a feeding/trapping corral and held in an individual cage, given water and fasted for approximately 12 hours prior to bleeding. The subjects were transferred to a squeeze-back cage and injected intramuscularly with ketamine 10 mg/kg body weight prior to venipuncture. Blood samples were collected from either the femoral or saphenous vein. Blood samples were forwarded to a commercial laboratory [Vetpath/Medpath Inc., Teterboro, New Jersey]. The hemograms including the hematocrits and erythrocyte indices were determined electronically on a Coulter counter model S.)

31 Bilimoria (1931) (Data from a series of 27 adult monkeys of both sexes of *Macaca mulatta* and *Macaca sinica*. The MCD was determined on a modified Young's eriometer as described by Pryce D. M. (1929), Lancet II, 275.)

32 Buchl et al. (1997) (Data were obtained from 527 healthy, domestically bred, and reared *Macaca mulatta* at the University of Texas, M. D. Anderson Cancer Center, Department of Veterinary Sciences, Bastrop, TX. The subjects included 251 nongravid females, 97 gravid females, and 179 males. The animals' pedigrees and birth dates were known. The blood samples were collected during each animal's annual physical examination. The monkeys were housed in a variety of settings: in harem groups, in peer groups, caged in pairs, and caged singly. Housing was principally outdoor in single-male, multi-female breeding groups, with infants being reared with their dams from 6 months to 1 year of age. Phlebotomies were performed at the femoral vein; ketamine was administered intramuscularly at a dosage of 5–10 mg/kg of body weight. The animals were acclimated to their surroundings and consequently, blood collection was conducted under familiar conditions. The assays were conducted at the on-site pathology department with a Baker System 9000 Hematology Analyzer. The values were grouped according to age, sex, and gravidity. Due to the absence of any specific indication, it is assumed that the subjects were not fasted prior to venisection. Superscript "a" indicates an increasing level and a significant difference in the Hct and Hb with the age of males 3–10 years old and nongravid females of breeding age (3 to greater than 10 yr). An asterisk signifies a significant difference between the sexes for a given parameter for a given age or subgroup.)

33 Chen et al. (2009) (Blood was collected from 18 Chinese 3–5 yr old Chinese *Macaca mulatta* of each sex [19 were 3 years old, 14 were 4 years old, and 3 were 5 years old]. The mean body weight was 4.7 kg. They had originated from wild monkeys in southwest China and were colony bred in Ping'an Animal Breeding Center in Chengdu, China. At the time of blood collection, the animal feeders slowly walked up to the subject, gently caught it with a net. The captured monkey's head and upper extremities were restrained to permit venous phlebotomy from a lower extremity. One female's blood sample was not assayed because of hemolysis. Anesthesia was not utilized, and the subjects were apparently not fasted. The hemogram including the erythrocyte indices and red cell distribution width coefficient of variability [%] were determined with the electronic instrument, Hematology XE-2100, Tokyo, Japan.)

(*Continued*)

TABLE 1   215

Table 1 (Continued)   Erythrocyte Counts and Related Values

	RBC	Hct	Hb	Diameter	MCV	MCH	MCHC

34 Fernie et al. (1994) (15 adult female *Macaca mulatta* were studied over a period of 3 years, 20 adult males were studied over a period of 6 years. The female rhesus monkeys were sampled about 25 times over the 3-year period while males were assayed two to six times over a 6-year period. The subjects were housed singly in stainless steel cages. The average estimated age was 11.4 years at the initiation of the study. In addition, nine infants [five males and four females] were assayed approximately 20 times from the age of 12 weeks and 122 weeks of age. The infants remained with their dams for 22 weeks, and then they were weaned and monitored for another 2 years. Infants 12–24 weeks of age were not given ketamine prior to phlebotomy. Conversely, older "infants" were sometimes administered ketamine and in other assays they were not sedated. [At weeks 42–122, ketamine was utilized except for weeks 46, 50, and 58]. The data for these periods are listed separately. Data derived from the analyses of infants given ketamine prior to venisection at 42, 54, and 62 weeks of age were compared with the results derived from the same five male and four female monkeys not given the drug at assays at 46, 50, and 58 weeks to assess the effects of administration of the drug. Values indicated by the superscript "x" signify a significant difference between the two sets of data for a given parameter. Blood samples were obtained by femoral puncture while the subject was lying on its back and restrained by a neck collar. Ketamine was administered to the males, and as indicated, for some of the older infant samplings. The samplings were conducted at 8:00–10:00 A.M. with the last feeding having been given the prior morning. The samples were considered as being obtained from a nonfasting subject as a monkey could have had access to any unconsumed provisions from the previous day's allotment or to food that could have fallen through the mesh flooring on to the litter pan. Hemograms were determined with a Coulter counter model S for the first 11 analyses of the females and the first two analyses of the males. The rest of the assays were derived using a Coulter counter model S-Plus IV. Reticulocyte counts were performed on supravitally stained smears of blood. The data were reported as median values.)

35 Allen et al. (1968) (25 adult female *Macaca mulatta* were hematologically evaluated prior to impregnation, during the gestation period, and following delivery. The monkeys were housed in individual cages. The phlebotomies were conducted while two attendants manually restrained a given monkey. Femoral vein aspirations were obtained prior to impregnation and at 30-day intervals for the first 90 days of pregnancy. From the 90th to the 150th day, blood was obtained every 10 days and then twice weekly until delivery. Following parturition, venipunctures were performed every 10 days for 40 days. The samplings totaled 10 ml of blood. Studies were performed at the Department of Pathology and the Regional Primate Research Center, University of Wisconsin. Hemoglobin levels were determined photometrically as cyanmethemoglobin, hematocrits were determined by capillary tube centrifugation, and the erythrocyte counts were derived with the use of an electronic particle counter. The red cell indices were calculated by the application of the Wintrobe formulas.)

36 Allen et al. (1966) (Captive pregnant rhesus monkeys were phlebotomized at the femoral vein at regular indicated intervals. The samplings totaled 5 mL of blood. Hemoglobin levels were determined photometrically as cyanmethemoglobin and hematocrits were determined by capillary tube centrifugation. The phlebotomies were conducted while two attendants manually restrained a given monkey. Studies were performed at the Department of Pathology and the Regional Primate Research Center. The average gestation was 166 days.)

37 Ponder et al. (1928) (Diameters of the erythrocytes were photomicrographically measured while in plasma or serum and in dry film smears; some of the data are also presented in Ponder et al. [1929].)

38 Hall (1929) (Captive laboratory monkeys fasted for 18–24 hours prior to sampling from the saphenous vein. Multiple samples were obtained over several months. Erythrocyte counts were performed with a Levy hemocytometer.)

(Continued)

TABLE 1

Table 1 (Continued) Erythrocyte Counts and Related Values

	RBC	Hct	Hb	Diameter	MCV	MCH	MCHC

39 Bicknese et al. (1993) (A study of 143 weanling subjects with weaning age of a mean 110 days. The study included 80 females and 63 males. The blood sampling was performed under manual restraint 1 week prior to actual weaning. Iron deficiency was defined as presenting an MCV $\leq$ 66 fL. Superscript "x" indicates a statistical difference for values of the indicated parameter between the iron-deficient and non-iron-deficient monkeys.)

40 Wolcott et al. (1973) (20 healthy young adult male rhesus monkeys aged 2–5 years were bled three times per week in a 30-day period. Ten mL blood per kg of body weight was withdrawn from the femoral vein or artery at each phlebotomy. Seven phlebotomies with assays were performed on each subject plus a final assay on day 30. The methodology for the hemograms was not described.)

42 Hassimoto et al. (2004) (The animals were 3.5 years old, males and females, bred in China. They were housed singly in stainless steel cages in a climate controlled room. The subjects were studied and "acclimated" for 6 months which included daily body touching, daily hand-to-hand feeding, 10 minutes of monkey chair restraint twice weekly and forced oral administration of water by use of a latex catheter twice per week and monthly blood samplings. The monkeys were fasted 16 hours prior to each sampling in the morning, once a month on the same day. The animals were restrained in a monkey chair for a cephalic vein phlebotomy. The hematologic values were determined electronically with a SF-3000, Sysmex, Kobe, Japan automatic analyzer. The data reported are mean values of 6 monthly analyses. The cumulative 6-month mean for the hematologic values was not statistically analyzed for differences between sexes. The erythrocytic indices were not reported but calculated for the table given earlier using the Wintrobe formulas. Only fleeting changes were observed during the 6-month long study of the erythrocyte count, hematocrit, and hemoglobin concentration.)

43 Horn et al. (1999) (The monkeys were male *Macaca mulatta* 3–5 years of age with a weight of 3.5–6 kg. They were housed singly or in pairs under standard environmental conditions indoors as part of a colony maintained by the Department of Laboratory Animal Resources, Merck and Co., Rahway, NJ. The monkeys were of Indian origin. The data were obtained from fasted, anesthetized animals given 8–10 mg of ketamine intramuscularly. They were fasted 16–24 hr. Fifty-three ketamine-anesthetized animals were studied over a 4-year period with an individual subject being analyzed 2–15 times. The results were averaged. The trachea was intubated, and an angiocath was inserted into the femoral artery proximal to the heart. In addition, an angiocath was inserted into the brachiocephalic vein to monitor the blood pressure for intravenous administration. Approximately 30 minutes after ketamine administration, a blood sample was taken from the arterial catheter. The hemograms were determined using a Bayer H1E blood analyzer. After the blood sample was taken another anesthetic agent, that is, pentobarbital, propofol, or isoflurane was then administered and 30 minutes later a second sample of blood was taken.)

44 Bennett et al. (1992) (The monkeys were adults, 11–22 years old healthy nonpregnant females, n = 15. They were housed singly and conditioned to handling. They were restrained by two handlers and appeared calm during the femoral bleeding. Blood samples were obtained before and approximately 15 minutes after intramuscular injection of 5–10 mg/kg of ketamine. Hematologic values were determined using the automated instrument Sysmex E-2500, Toa Medical Electronics, Kobe, Japan. Pre- and post-ketamine values for each variable were compared by a paired-difference t-test. Superscript "x" indicates a statistical difference for the values of the indicated parameter.)

(Continued)

TABLE 1 217

Table 1 (*Continued*) Erythrocyte Counts and Related Values

	RBC	Hct	Hb	Diameter	MCV	MCH	MCHC

[45] Loomis et al. (1980) (Nonpregnant, adult, female rhesus monkeys were randomly chosen from a group housed indoors in individual squeeze-back cages. Studies were conducted at the California Primate Research Center, University of California at Davis. When ketamine was administered, the monkey was immobilized with the squeeze mechanism of its home cage, a limb was manipulated through the bars of the cage and ketamine 15 mg/kg body weight was administered intramuscularly. Simultaneously, blood was obtained from either the cephalic or saphenous vein. Additional samples were obtained 10 and 20 minutes later. Non-ketamine samples were obtained from the same monkeys with the same physical maneuvering using a limb manipulated through the bars of the squeeze-back cage. Two weeks elapsed between the time an individual monkey was immobilized with ketamine and the time it was used as a control. No indication was given whether the monkeys were fasting. The hematocrits and leukocyte counts were determined with a Coulter cell counter. Hb concentrations, erythrocyte counts, and MCV were established but not reported. The superscript "x" indicates a significant difference in the hematocrits between the ketamine-treated, restrained subjects and the control, restrained monkeys not administered ketamine for the indicated time interval.)

[46] Vondruska et al. (1974) (The subjects were housed in steel cages at BIO-TEST Laboratories, Northbrook, IL. The monkeys were juveniles weighing 2–3 kg and were studied over a period of 14 months. The subjects were fasted 18 hr prior to phlebotomy. They were manually restrained at venisection. The hemograms were electronically determined using a Coulter counter.)

[47] Rogers et al. (2005) (All monkeys were born at the Tulane National Primate Research Center, LA and were housed individually. Normal healthy pregnant females were used. The gestational age and the delivery dates were calculated during early gestation by utilizing timed breeding dates and ultrasound measurements. The infants were delivered by cesarean section on gestational day 155 prior to onset of labor [average gestation for the rhesus monkey is 165 days]. Nine of the dams were 4–5 years old and first time pregnant, while eight were 5–6 years old and second time pregnant. The dams were pre-anesthetized with glycopyrrolate followed by ketamine sedation and isoflurane gas anesthesia. At cesarean section, the uterus was exteriorized, the infant was removed from the amniotic sac, and the umbilical cord was clamped and severed. A venous sample of blood was obtained by venipuncture from the segment of the umbilical cord attached to the placenta prior to the removal of the placenta from the uterus. The analyses were conducted on a Coulter counter T540 or a Bayer ADVIA 120 Hematology System; the latter instrument was normalized against the Coulter counter instrument.)

[48] Lewis (1977) (13 adult [>4 years old] monkeys were studied at the Primate Research Laboratory, Department of Physiology, University of Pittsburgh School of Medicine. Six were males and seven were females. Three of the males and four females were analyzed by manual hematologic methods while three males and three females were studied using the electronic techniques of the Coulter counter S. The mean corpuscular diameter was determined by examination of Wright-stained blood smears.)

[49] Zeng et al. (2011) (Subjects were obtained from Hainan Jingang Laboratory Co., Haikou, China. The monkeys were housed singly in steel cages in an animal room with a controlled environment. All subjects were 4–5 years old, the six males weighed 3.1–6.0 kg while the five females weighed 2.8–3.6 kg. Water was available *ad libitum*. Samples were obtained circa 1400 and 1500 hr from the cephalic vein from conscious monkeys. Blood samples were obtained after the monkeys were fasted for 0, 8, 16, and 24 hr. The samples were obtained at least 24 hr after a prior sample had been obtained. Each subject served as its own control. The data were statistically analyzed. The superscript "x" indicates a significant difference from the non-fasting, zero time level for the indicated parameter and fasting interval. Statistical analyses of male versus female values were not conducted. The analyses were conducted on a Medonic CA620 VET hematology analyzer, Boule Medical, Stockholm, Sweden.)

(*Continued*)

Table 1 (*Continued*) Erythrocyte Counts and Related Values

	RBC	Hct	Hb	Diameter	MCV	MCH	MCHC

50 Martin et al. (1973) (All monkeys were conceived in the laboratory and were born within two standard deviations of the mean length of gestation in this laboratory [164 days]. They were separated from their mothers within a few hours of birth and were hand-reared on a commercial human formula. The blood samples were obtained by femoral phlebotomy and the hematocrits were determined by the microhematocrit technique. To eliminate abnormal values, hematocrits lying outside 2 SD were deleted and the mean was recomputed. Thus although 170 subjects were studied, the "n" for any given analysis varies. The monkeys were studied from birth through 2 years of age. Analyses were performed once during the first 3 days of life; at 1, 2, 3, 4, and 8 weeks; and monthly during the rest of the year. The analyses were conducted every 6 months during the second year. The reported data were not separated by sex.)

52 Lugo-Roman et al. (2010) (The subjects were six male and two female adult *Macaca mulatta* with a weight range of 5.5–7 kg and an estimated age of 5–5.5 years. They were maintained in single cages at the Walter Reed Army Institute of Research. The monkeys were fasted overnight and administered ketamine intramuscularly 10 mg/kg of body weight on each of three consecutive days prior to venisection. The blood analyses were conducted electronically using an Advia 120 Hematology System, Siemens Health Care Diagnostics.)

53 Kim et al. (2005a) (The subjects were five male *Macaca fascicularis* 3–4 years of age, mean weight 2.52 kg. They had been transported from Japan to the laboratory in Korea. The total transit time was 15 hr [2.5 hr in an air-conditioned truck, 3 hr in a plane, 4 hr in an air-conditioned truck in Korea, and 5.5 hr holding time at the airport before and after the flight]. The monkeys were assayed in a quarantine room on arrival and subsequently weekly during the following 35 days. Blood samples were obtained per cephalic vein and within 30 minutes of administration of ketamine. The monkeys were housed individually in stainless steel cages. The before arrival and arrival weights along with the subsequent weights suggest that the subjects became dehydrated during the journey. The mean values of a given parameter with different superscripts are significantly different. The hematological analyses were conducted using the autoanalyzer ADVIA Hematology System Bayer.)

54 Lee et al. (2013) (The subjects were female, untrained monkeys 3–4 years of age. They were imported from China, accepted in South Korea, and quarantined for 1 month. They were then acclimated to the animal facility for about 6 months. They were maintained singly in cages. The study involved training the subjects to accommodate to a monkey chair under stress-free conditions. Baseline values were collected after instant immobilization using the squeeze device of the cage without anesthesia. Monkey chair training was conducted daily, 5 days/week for 1 month. The training time was 30 minutes during which the monkeys were trained to submit to simple procedures as examinations, sampling, and drug administration. Blood samples were collected weekly for hematologic studies and cortisol analyses. Hematologic analyses were conducted on an automatic analyzer MS9-5V, Melet Scholesing Lab Co. There were no significant differences in the erythrogramic values at any point in time except for the hematocrit during chair training in comparison with baseline values. Different letters, that is, superscripts "a" and "b," indicate that the corresponding values are statistically different.)

55 Winkle (1951) (Laboratory-housed *Macaca mulatta* at the University of Minnesota. The subjects were 10–15 months old with average weights of 1.5–2.5 kg. The erythrocyte counts were performed manually using a Levy-Hausser counting chamber. Hayem's diluting fluid was used. Hematocrits were determined by using a Wintrobe hematocrit tube with centrifuging for 30 minutes. Hemoglobin concentration was established by the use of the Evelyn photoelectric colorimeter and quantitated as oxyhemoglobin. Reticulocyte counts were determined on direct blood smears. Microscope slides were previously prepared with brilliant cresyl blue stain and blood was thinly smeared on them. The erythrocytes were supravitally stained.)

(Continued)

TABLE 1 219

Table 1 (*Continued*) Erythrocyte Counts and Related Values

	RBC	Hct	Hb	Diameter	MCV	MCH	MCHC

56 Smucny et al. (2001, 2004) (Aggregated data of the Primate Aging Database, a multicentered data base. The animals were maintained at three institutions (1) the National Institute of Aging research group at the National Institutes of Health Animal Center, Poolesville, MD, (2) the Wisconsin Regional Primate Research Center, Madison, WI, and (3) the Oregon Regional Primate Research Center, Beaverton, OR. All data were obtained from healthy, primarily indoor housed animals that were regularly sampled at health assessments over several years. All participants had known dates of birth. Data were analyzed from subjects 6–36 years old. Blood samples were drawn either without sedation or following ketamine injection (7–10 mg, IM) in the morning after an overnight fast. The long-term mean erythrograms were based on a population of 231–246 monkeys with 2952–3093 samplings for each erythroid parameter. Analyses were conducted at independent clinical laboratories in Maryland, Oregon, and Wisconsin. Extensive quality control measures were performed at each facility to insure intra-facility standardization.)

57 Ibanez-Contreras et al. (2013) (Subjects ranged between 4 and 6 years of age with an average weight of 6 kg. Fourteen pregnant *M. mulatta* in their first trimester were blood sampled between the 6th and 10th week of pregnancy [day 56 ± 14]. Fourteen nonpregnant females of the same age range were sampled 5 days after ovulation. The subjects were maintained at the Research Center Proyecto CAMINA A.C., Mexico City C.P. 14050. They were isolated from their group and placed in individual cages and had an 8-hr fasting period prior to phlebotomy. They were administered ketamine 10 mg/kg IM and xylazine 0.5 mg/kg IM before venisection. After recovery post-venisection they were returned to their group. Data were subjected to statistical analysis.)

58 Switzer et al. (1970) (These are studies of *Macaca mulatta* maintained in a laboratory breeding colony. Each animal had spent at least 6 months in quarantine and conditioning. They were housed individually except for the 72-hr encounter with one male for breeding. One group of 20 subjects was monitored from the third trimester of pregnancy and postpartum for a period of 8 weeks. Normal animals used for comparison were more than 60 days postpartum. Another group of 200 pregnant monkeys were followed on a trimester basis during pregnancy and postpartum for 8 weeks. Blood was obtained by femoral venipuncture. The RBC counts were performed with a Coulter counter. Hb was quantitated spectrophotometrically as cyanmethemoglobin, and the Hct was determined per the centrifuged microhematocrit technique. Reticulocyte counts and erythrocyte sedimentation rates were also established.)

60 Bonfanti et al. (2009) (60 cynomolgus monkeys were randomly selected on the basis of body weight and breeding origin and assigned to two groups each of 15 males and 15 females. One group was comprised of purpose bred animals about 2 years old (i.e., juveniles). The other group was comprised of captured animals of similar age obtained from a supplier located in Mauritius. All subjects were housed at the primate facility and underwent a quarantine period of at least 40 days. The animals had free access to food and were fasted overnight before blood sampling. Conscious animals were restrained by experienced animal technicians and blood was obtained by femoral phlebotomy. Hematology analyses were performed using a Technicon H1E system.)

61 Takenaka (1981, 1986) (The subjects were wild [young adults and mature adults as determined by dentition]. They were caught with a nylon net, administered ketamine, sampled, and released. They were members of five groups residing on Bali Island, that is, Teluk Terima, Pulaki, Sangeh, Ubud, and Kukuh. Another group residing in Gunung Meru, West Sumatra, Indonesia was also studied. The erythrocytes were enumerated by manual microscopy (counting chamber), and the hematocrits were determined by centrifugation. The hemoglobin concentration was quantitated as cyanmethemoglobin. For the determination of hemoglobin content of blood, the blood was diluted with physiologic saline, stored in ice, brought to the laboratory having a spectrometer, and analyzed within a week of collection.)

(Continued)

Table 1 (*Continued*)   Erythrocyte Counts and Related Values

	RBC	Hct	Hb	Diameter	MCV	MCH	MCHC

62 Bourges-Abella et al. (2014) (Specimens from 272 healthy cynomolgus monkeys maintained at BioPRIM, a primatology center at Baziege, France. The subjects originated mainly from breeding facilities in Mauritius, Philippines, and Vietnam. The purpose of the study was to establish hematologic reference levels for this species according to the International Federation of Clinical Chemistry and Clinical and Laboratory Standards Institute guidelines. The monkeys were housed in collective cages. Most of the animals were fasted overnight; they were administered ketamine and had a femoral phlebotomy. The studied group was composed of 183 males (67%) and 89 females (33%). Most animals were born in captivity (231/172 = 85%). Twenty-three females and 18 males were captured, and their age was unknown. The ages of the remaining purpose-bred animals ranged 17.3–118.4 months but most of the animals (60%) were 2- to 3-year-old juveniles. Analyses were performed with a Sysmex KX-21 analyzer. The data were rigorously statistically analyzed. Sex had no significant effect on the erythrogramic values including the erythrocyte indices.)

63 Schuurman et al. (2005) (Subjects originated from local stock at Covance Laboratories, Madison, WI. Determinations were conducted on Abbott Cell-Dyn 3500 and Beckman Coulter S-plus IV instruments. A total of 53 determinations (samplings) were conducted for the males as well as for the females.)

64 Verlangieri et al. (1985) (A comparison of erythrograms of male *Macaca fascicularis* and male *Macaca arctoides*. The *Macaca fascicularis* were adults approximately 8–15 years old and weighed 4.0–7.5 kg while the *Macaca arctoides* were adults approximately 8–10 years of age and weighed 10—15 kg. The subjects were housed individually and had been at the laboratory for a minimum of 6 months. Superscript "x" indicates a significant difference between *M. fascicularis* and *M. arctoides*. The animals were fasted at least 17 hr and administered 10 mg ketamine/kg IM prior to phlebotomy at the femoral triangle [obtaining venous or arterial blood]. Determinations were conducted electronically.)

65 Wang et al. (2012) (Healthy, singly caged juvenile *Macaca fascicularis* monkeys 2.5–3.5 years old were studied. Blood was obtained from nonanesthetized subjects after fasting for 16 hr. Samples were obtained from the femoral vein. Analyses were performed electronically using a Sysmex Hemacytometer XT-2000i. Reticulocyte counts were determined.)

66 Yoshida et al. (1986b) (Study of effects of ketamine anesthesia on female *M. fascicularis*; dosage 15 mg/kg body weight. Controls were injected with saline. The subjects were born in the wild. The mean weight was 3.5 kg, and the monkeys were thus assumed to be adults. Determinations were conducted electronically. The erythrocyte indices were determined from the reported data by the application of the Wintrobe formulas.)

67 Xie et al. (2013) (The monkeys were maintained at the Xishan Zhongke Laboratory Animals Co, Suzhou, China. All subjects were housed in same-sex social groups that were formed when the monkeys were 6 months of age. They were housed in indoor pens. After overnight fasting (14–16 hr), conscious monkeys were restrained humanely by experienced animal caretakers. Blood was drawn from the cephalic vein. The ages of the monkeys ranged from 13 to 72 months. A total of 374 males and 547 females were analyzed. Analyses were performed using a Sysmex XT-2000iv automated analyzer. Extensive statistical analyses were conducted.)

68 Koga et al. (2005) (95 male and 95 female Chinese–bred *Macaca fascicularis* monkeys 3–7 years old were studied. Analyses were performed on femoral blood utilizing the automated cell counter ADVIA 120, Bayer Medical [dual angle laser flow cytometric measurement]. A total of 220 samplings were analyzed for each parameter for each sex. All analyses were conducted four times a month.)

(Continued)

TABLE 1   221

Table 1 (*Continued*) Erythrocyte Counts and Related Values

	RBC	Hct	Hb	Diameter	MCV	MCH	MCHC

69 Yoshida et al. (1986a) (Blood samples were collected from more than 1000 *Macaca fascicularis* that were bred and reared under uniform environmental conditions. They were reared using indoor individual cages at the Tsukuba Primate Center for Medical Science. The ages of the subjects are specific. The hematologic values were determined by electronic methods including the hematocrits. It was not determined whether or not ketamine anesthesia was used during the phlebotomies.)

70 Santiyanont et al. (1977) (Young, individually caged *Macaca fascicularis* were divided into two groups apparently three subjects per group. The members of one cohort were placed of a vitamin E–deficient diet, and the constituents of the other cohort were placed on the same diet plus a supplement of 80 mg DL–α–tocopherol acetate per day. Femoral blood samples were obtained at 3-month intervals. The duration of the experiment was 18 months. Bone marrow was aspirated from the posterior iliac crest while the animals were anesthetized with chlorpromazine. The superscript "a" indicates a significant difference between the values for the indicated parameters of normal and vitamin E–deficient monkeys. The techniques utilized in establishing the hemograms were not cited.)

71 From Charles River Laboratories, cited by Bernacky et al. (2002) (*Macaca fascicularis*, specific age and other factors were not cited although it is assumed the subjects were mature adults and the females were nongravid. Modern electronic techniques including electronic determination of MCV and Hct are assumed. Presence or absence of anesthesia was not indicated. N = 80 but the distribution between males and females is not indicated.)

72 Fujiwara et al. (1974) (The monkeys used in this experiment had been kept at this facility for more than 2 yr. Their ages were estimated at more than 5 years based on dental examination. The subjects were individually housed. Twenty-six became pregnant following a one-to-one mating program in a cage for 7 days that started on the 11th day of the menstrual cycle. Nine other subjects did not become pregnant and were used as nonpregnant controls. Fourteen pregnant monkeys were utilized for erythrogramic studies. A venipuncture was performed every other week during pregnancy beginning from the first day of mating and 1, 2, 4, and 8 weeks after parturition. The remaining 12 pregnant and 9 nongravid monkeys were used for erythrocyte sedimentation analyses and other pre- and postdelivery hemogramic and biochemical studies. The analyses were conducted by manual red cell counting, hematocrits by centrifugation of capillary tubes, and hemoglobin determinations by the Sahli technique.)

73 Adams et al. (2014) (Laboratory-housed, mature cynomolgus monkeys at the Biologic Resources Laboratory, University of Illinois at Chicago. The samples were obtained over a 3-year period and were apparently obtained form ketaminized subjects (10 mg/kg IM). Analyses were performed on an automated hematology analyzer Advia 120, Siemens Healthcare Diagnostics, Tarrytown, NY.)

74 Wolford et al. (1986) (Subjects were maintained at the Medical Research Division, American Cyanamid Company, Pearl River, NY. Blood was obtained from the femoral vein. The animals were conscious and they were of all ages. Subjects were obtained from Primate Imports Corporation, Port Washington, NY. They had been in the wild and thus exact ages were not available. The analyses were conducted with a Coulter S-Plus.)

(Continued)

Table 1 (*Continued*) Erythrocyte Counts and Related Values

	RBC	Hct	Hb	Diameter	MCV	MCH	MCHC

[75] Giulietti et al. (1991) (The subjects were members of a colony at the Instituto Superiore di Sanita, Rome, Italy. Most of the monkeys had been born in the colony; the only imported animals were four adults that were more than 10 years old and had been housed in the colony for >10 years. The group was comprised of 66 subjects: 35 [2-month to 15-year-old males weighing a mean 3.2 kg] and 31 [1.5-months to 8-year-old females weighting a mean 3.2 kg]. Pregnant females were excluded from the study. The colony was caged in an indoor–outdoor enclosure. They were thus housed as a group and not individually caged. Each subject was captured and sedated with 10 mg ketamine at the time of analysis. The monkeys were not given food on the day of sampling and blood was obtained from a femoral vein or artery between 8:00 and 10:00 A.M. The hemogramic analyses were performed electronically on a Hycel Data 8 Hematology Analyzer. The data were statistically analyzed for differences between the sexes.)

[76] Matsuzawa et al. (1993) (The data were collected from 67-member companies of the Japan Pharmaceutical Manufacturers Association covering a population of ~540 male and 500 female wild-caught cynomolgus monkeys between 2 and 5 years old. The erythrocyte counts were believed to be determined by electronic means, the microhematocrit technique was employed at a few facilities, and the hemoglobin level was established most often as cyanmethemoglobin and as oxyhemoglobin at other institutions.)

[77] Golub et al. (1999) (The subjects were members of the colony maintained at the California Regional Primate Research Center, University of California, Davis. They were adolescent females with an average age of 29.6 months. The subjects were maintained on a normal diet or one that was mildly deficient in iron and zinc. Hemograms were conducted when the experiment was initiated, at the end of the 3 months of study and thereafter at monthly intervals during the next 3 months. The hemograms were performed at the cited research center and were therefore assumed to have been electronically assayed. Superscripts indicate a significant difference between the values of a given parameter of an indicated test group of monkeys.)

[78] Munro (1987) conducted at the University of Montana Primate Laboratory. Hematocrits were determined by centrifugation and the other hematologic determinations are presumed to be electronically derived as they were determined by the Laboratory of Western Montana Clinic.

Alphabetical Index of Investigators' Reports Listed in Table 1
(Superscripts Are Those Utilized in Table 1)

Adams et al.[73] (2014)
Allen et al.[35] (1968)
Allen et al.[36] (1966)
Andrade et al.[1] (2004)
Bennett et al.[44] (1992)
Bicknese et al.[39] (1993)
Bilimoria[31] (1931)
Bonfanti et al.[60] (2009)
Bourges-Abella et al.[62] (2014)
Buchl et al.[32] (1997)
Charles River Laboratories[71] cited by Fox et al. (2002)
Chen et al.[22] (2002)
Chen et al.[33] (2009)
Fernie et al.[34] (1994)
Fujiwara et al.[72] (1974)
Giuletti et al.[75] (1991)
Golub et al.[77] (1999)
Hall[38] (1929)
Hassimoto et al.[42] (2004)
Hom et al.[43] (1999)
Ibanez-Contreras et al.[57] (2013)
Ives et al.[23] (1956)
Kessler et al.[25] (1983a)
Kessler et al.[30] (1983b)
Kim et al.[53] (2005a)
Kim et al.[12], (2005b)
King and Gargus[2] (1967)
Koga et al.[68] (2005)
Kupper et al.[29] (1976)
Lee et al.[54] (2013)
Lewis[48] (1977)
Lilly et al.[24] (1999)
Loomis et al.[45] (1980)
Lugo-Roman et al.[52] (2010)
Martin et al.[50] (1973)
Matsumoto et al.[8] (1980)
Matsuzawa et al.[76] (1993)
Matsuzawa et al.[13] (1994)

References

Adams C. R., Halliday L. C., Nunamaker E. A., and Fortman J. D. (2014). Effects of weekly blood collection in male and female cynomolgus macaques (*Macaca fascicularis*). *J. Am. Assoc. Lab. Anim. Sci.* 53, 81–88.

Ageyama N., Shibata H., Narita H., Hanari K. et al (2001). Specific gravity of whole blood in cynomolgus monkeys (*Macaca fascicularis*), squirrel monkeys (*Saimiri sciureus*), and tamarins (*Saguinus labiatus*) and total blood volume in cynomolgus monkeys. *Contemp Top Lab Anim Sci.* 40(3), 33–35.

Alberts B., Bray D., Lewis J., Raff M. et al. (1983). Cell growth and divison. Section: Cell division. In: *The Molecular Biology of the Cell*, 1st edn., Chapter 11. Garlane Publishing Inc., New York, pp. 646–668.

Allen J. R. and Ahlgren S. A. (1968). A comparative study of the hematologic changes in pregnancy in the *Macaca mulatta* monkey and the human female. *Am. J. Obst. Gynecol.* 100, 894–903.

Allen J. R. and Siegfried L. M. (1966). Hematologic alterations in pregnant rhesus monkeys. *Lab. Anim. Care* 16, 465–471.

Ameri M., Boulay M., and Honor D. J. (2010). What is your diagnosis? Blood smear from a cynomolgus monkey (*Macaca fascicularis*). *Vet. Clin. Pathol.* 39, 257–258.

Anderson J. H., Keen C. L., Lonnerdal B., Leninger R. et al. 1983. Iron deficiency in outdoor corral housed juvenile rhesus monkeys. *Lab. Anim. Sci.* 33, 494.

Andrade M. C. R., Ribeiro C. T., Ferreira da Silva V., Molinaro E. M., Goncalves M. A. B., Marques M. A. P., Cabello P. H., and Leite J. P. G. (2004). Biologic data of *Macaca mulatta*, *Macaca fascicularis* and *Saimiri sciureus* used for research at the Fiocruz primate center. *Mem. Inst. Oswaldo Cruz* 99, 581–589.

Ausman L. M. and Hayes K. C. (1974). Vitamin E deficiency anemia in old and new world monkeys. *Am. J. Clin. Nutr.* 27, 1141–1151.

Bender M. A. (1955). Blood volume of the rhesus monkey. *Science* 122, 156.

Bennett J. S., Gossett K. A., McCarthy M. P., and Simpson E. D. (1992). Effects of ketamine hydrochloride on serum biochemical and hematologic variables in rhesus monkeys (*Macaca mulatta*). *Vet. Clin. Pathol.* 21, 15–18.

Bentson K. L., Capitanio J. P., and Mendoza S. P. (2003). Cortisol responses to immobilization with Telazol or ketamine in baboons (*Papio cynocephalus*/anubis) and rhesus macaques (*Macaca mulatta*). *J. Med. Primatol.* 32, 148–160.

Bernacky B. J., Gibson S. V., Keeling M. E., and Abee C. R. (2002). Nonhuman primates. In: Fox J. G., Anderson L. C., Loew F. M., and Quimby F. W. (eds.), *Laboratory Animal Medicine*, 2nd edn., Chapter 16. Academic Press, Elsevier Science, New York, pp. 675–791.

Bessis M. (1958). L'îlot érythroblastique unité fonctionnelle de la moelle osseuse. *Rev. Hematol.* 13, 8–11.

Bessis M. (circa 1960). *Life and Death of Red Cells* (a 16 mm film with dialogue). National Blood Transfusion Center, Sandoz Film Library, Sandoz Pharmaceuticals, Hanover, NJ; Sandoz Ltd., Basel, Switzerland.

Bessis M. (1966). *The Life Cycle of the Erythrocyte*, Sandoz Monographs. Sandoz, Basel, Switzerland, 108pp.

Bessis M. (1973). The erythrocytic series. In: *Living Blood Cells and their Ultrastructure*, Chapter II. Springer-Verlag, Berlin, Germany, pp. 86–89.

Bessis M. (1977). Erythrocytic series. In: *Blood Smears Reinterpreted*, Chapter 2. Springer-Verlag, Berlin, Germany, pp. 38–43.

Beutler E. (1977). Osmotic fragility. In: Williams W. J., Beutler E., Erselev A. J., and Rundles R. W. (eds.), *Hematology*, 2nd edn., Chapter A12. McGraw-Hill Book Company, New York, pp. 1609–1610.

Bicknese E. J., George J. W., Hird D. W., Paul-Murphy J. et al. (1993). Prevalence and risk factors for iron deficiency in weanling rhesus macaques. *Lab. Anim. Sci.* 43, 434–438.

Bilimoria H. S. (1931). Blood findings in normal monkeys. *Indian J. Med. Res.* 19, 431–432.

Bohm R. P. Jr., Martin L. N., Davidson-Fairburn B., Baskin G. B. et al. (1993). Neonatal disease induced by SIV infection of the rhesus monkey (*Macaca mulatta*). *Aids Res. Hum. Retroviruses* 9, 1131–1137.

Bonfanti U., Lamparelli D., Colombo P., and Bernardi C. (2009). Hematology and serum chemistry parameters in juvenile cynomolgus monkeys (*Macaca fascicularis*) of Mauritius origin: Comparison between purpose-bred and captured animals. *J. Med. Primatol.* 38, 228–235.

Boonjawat J., Wilairat P., and Vimokesant S. L. (1979). Alteration in bone marrow RNA of vitamin E-deficient monkey, *Macaca fascicularis*. *Am. J. Clin. Nutr.* 32, 2065–2067.

Bourges-Abella N., Geffre A., Moureaux E., Vincenti M. et al. (2014). Hematologic reference intervals in cynomolgus (*Macaca fascicularis*) monkeys. *J. Med. Primatol.* 43, 1–10. doi: 10.1111/jmp.12077.

Bourne G. H. (1975). Collected anatomical and physiological data from the rhesus monkey. In: Bourne G. H. (ed.), *The Rhesus Monkey*, Vol. I: *Anatomy and Physiology*. Academic Press, New York, pp. 1–63.

Braun S. E., Walker E., Haupt E. M., Penney T. P. et al. (2013). Increased granulocyte/macrophage progenitor activity leads to thrombocytopenia and anemia in SIV-infected rhesus macaque. *J. Med. Primatol.* 42, 254 (abst #2).

Buchl S. J. and Howard B. (1997). Hematologic and serum biochemical and electrolyte values in clinically normal domestically bred rhesus monkeys (*Macaca mulatta*) according to age, sex and gravidity. *Lab. Anim. Sci.* 47, 528–533.

Capitanio J. P., Kyes R. C., and Fairbanks L. A. (2006). Considerations in the selection and conditioning of old world monkeys for laboratory research: Animals from domestic sources. *ILAR J.* 47, 294–306.

Castro M. I., Rose J., Green W., Lehner N. et al. (1981). Ketamine-HCl as a suitable anesthetic for endocrine, metabolic, and cardiovascular studies in *Macaca fascicularis* monkeys. *Proc. Soc. Exp. Biol. Med.* 168, 389–394.

Chang C., Lee I., Fletcher M. D., and Tarantal A. F. (2005). Effect of age on the frequency, cell cycle, and lineage maturation of rhesus monkey (*Macaca mulatta*) CD34+ and hematopoietic progenitor cells. *Pediatr. Res.* 58, 315–322.

Chasis J. A. and Mohandas N. (2008). Erythroblastic islands: Niches for erythropoiesis. *Blood* 112, 470–478.

Chen Y., Ono F., Yoshida T., and Yoshikawa Y. (2002). Relationship between body weight and hematological and serum biochemical parameters in female cynomolgus monkeys (*Macaca fascicularis*). *Exp. Anim.* 51, 125–131.

Chen Y., Qin S., Ding Y., Wei L. et al. (2009). Reference values of clinical chemistry and hematology parameters in rhesus monkeys (*Macaca mulatta*). *Xenotransplantation* 16, 496–501.

Chi Z., Xiao-Xiao W., Lu W., Ying X. et al. (2010). Application of flow cytometry to detect ABO blood group antibody levels in rhesus monkeys and cynomolgus monkeys. *Zool. Res.* 32, 56–61.

Clarke A. S. (1986a). Species differences among macaques in physiological responses to handling, novelty and restraint stress. *Am. J. Primatol.* 10, 394 (abstract).

Clarke A. S. (1986b). Species differences in fearful and aggressive behavior among macaques. *Am. J. Primatol.* 10, 395 (abstract).

Clarke M. R., Phillippi K. M., Falkenstein J. A., Moran E. A. et al. (1990). Training corral-living rhesus monkeys for fecal and blood sample collection. *Am. J. Primatol.* 20, 181 (abstract).

Cohen B. S. (1953). Bone marrow aspiration in the monkey (*Macacus rhesus*). *Blood* 8, 661–663.

Coleman K., Pranger L., Maier A., Lambeth S. P. et al. (2008). Training rhesus macaques for venipuncture using reinforcement techniques: A comparison with chimpanzees. *J. Am. Assoc. Lab. Anim. Sci.* 47, 37–41.

Davies H. G. (1961). Structure in nucleated erythrocytes. *J. Biophys. Biochem. Cytol.* 9, 671–687.

Dinning J. S. and Day P. L. (1957). Vitamin E deficiency in the monkey. I. Muscular dystrophy, hematologic changes, and the excretion of urinary nitrogenous constituents. *J. Exp. Med.* 105, 395–402.

Donahue E. R., Wang E. A., Stone D. K., Kamen R. et al. (1986). Stimulation of haematopoiesis in primates by continuous infusion of recombinant human GM-CSF. *Nature* 321, 872–875.

Elvidge H., Challis J. R. G., Robinson J. S., Roper C. et al. (1976). Influence of handling and sedation on plasma cortisol in rhesus monkeys (*Macaca mulatta*). *J. Endocrinol.* 70, 325–326.

Emerson C. L., Tsai C.-C., Holland C. J., and Diluzio M. E. (1990). Recrudescence of Entopolypoides macaci Mayer, 1933 (Babisiidae) infection secondary to stress in long-tailed macaques (*Macaca fascicularis*). *Lab. Anim. Sci.* 40, 169–171.

Fernie S., Wrenshall E., Malcolm S., Bryce F. et al. (1994). Normative hematologic and serum biochemical values for adult and infant rhesus monkeys (*Macaca mulatta*) in a controlled laboratory environment. *J. Toxicol. Environ. Health* 42, 53–72.

Fitch C. D., Broun G. O. Jr., Chou A. C., and Gallagher N. I. (1980). Abnormal erythropoiesis in vitamin E-deficient monkeys. *Am. J. Clin. Nutr.* 33, 1251–1258.

Fortman J. D., Hewett T. A., and Bennett B. T. (2002). Important biological features. In: *The Laboratory Nonhuman Primate*, Chapter 1. CRC Press, Boca Raton, FL, pp. 1–34.

Forsyth R. P., Nies A. S., Wyler F., Neutze J. et al. (1968). Normal distribution of cardiac output in the unanesthetized, restrained rhesus monkey. *J. Appl. Physiol.* 25, 736–741.

Fujiwara T., Suzaki Y., Yoshioka Y., and Honjo S. (1974). Hematological changes during pregnancy and postpartum period in cynomolgus monkeys (*Macaca fasicularis*). *Jikken Dobutsu. Exp. Anim.* 23, 137–146.

Fuller G. B., Hobson W. C., Reyes F. I., Winter J. S. D. et al. (1984). Influence of restraint and ketamine anesthesia on adrenal steroids, progesterone, and gonadotropins in rhesus monkeys. *Proc. Soc. Exp. Biol. Med.* 175, 487–490.

Gill A. F., Ahsan M. H., Lackner A. A., and Veazey R. S. (2012). Hematologic abnormalities associated with simian immunodeficiency virus (SIV) infection mimic those in HIV infection. *J. Med. Primatol.* 41, 214–224.

Giulietti M., La Torre R., Pace M., Iale E. et al. (1991). Reference blood values of iron metabolism in cynomolgus macaques. *Lab. Anim. Sci.* 41, 606–608.

Gleason N. N. and Wolf R. E. (1974). *Entopolypoides macaci* (babesiidae) in *Macaca mulatta*. *J. Parasitol.* 60, 844–847.

Glomski C. A., Chao C.-F., and Zuckerman G. B. (1982). Haemolytic anaemia in rhesus monkeys induced by methylcellulose. *Lab. Anim.* 16, 310–313.

Glomski C. A., Hagle R. E., and Pillay S. K. (1971a). Survival of chromium-51-labeled erythrocytes in the rhesus monkey. *Am. J. Vet. Res.* 32, 149–154.

Glomski C. A. and Pica A. (2006). *Erythrocytes of the Poikilotherms: A Phylogenetic Odyssey.* Fowell and Davies Ltd., London, UK.

Glomski C. A. and Pica A. (2011). *The Avian Erythrocyte: Its Phylogenetic Odyssey.* Science Publishers/CRC Press/Taylor & Francis Group, Boca Raton, FL. Online: http://bookzz.org.

Glomski C. A., Pillay S. K., and Hagle R. E. (1971b). Survival of [50]Cr-labeled erythrocytes as studied by instrumental activation analysis. *J. Nucl. Med.* 12, 31–34.

Glomski C. A., Pillay K. K. S., and Macdougall L. G. (1976). Erythrocyte survival in children as studied by labeling with stable [50]Cr. *Am. J. Dis. Child.* 130, 1228–1230.

Golub M. S., Keen C. L., and Gershwin M. E. (1999). Behavioral and hematologic consequences of marginal iron-zinc nutrition in adolescent monkeys and the effect of a powdered beef supplement. *Am. J. Clin. Nutr.* 70, 1059–1068.

Gregersen M. I., Sear H., Rawson R. A., Chien S. et al. (1959). Cell volume, plasma volume, total blood volume and F_{cells} factor in the rhesus monkey. *Am. J. Physiol.* 196, 184–187.

Gulliver G. (January 1845). On the size of the red corpuscles of the blood in the vertebrata. *Proc. Zool. Soc. Lond.* 13(1), 93–102.

Gulliver G. (1875). Observations on the sizes and shapes of the red corpuscles of the blood of vertebrates, with drawings of them to a uniform scale, and extended and revised tables of measurements. *Proc. Zool. Soc. Lond.* 31, 474–495.

Guthkelch A. N. and Zuckerman S. (1937). The red cell count of macaques in relation to the menstrual cycle. *J. Physiol.* 91, 269–278.

Hall B. E. (1929). The morphology of the cellular elements of the blood of the monkey, *Macacus rhesus. Folia Haematol.* 38, 30–43 plus one plate.

Hansen V. K. and Wingstrand K. G. (1960). Further studies on the non-nucleated erythrocytes of *Maurolicus mulleri* and comparisons with the blood cells of related fishes. In: *Carlsberg Foundation Oceanographic Expedition*, Dana Report Number 54, Vol. X. A. R. Host and Son, Copenhagen, Denmark, pp. 1–15.

Harlow C. M. and Selye H. (1937). The blood picture in the alarm reaction. *Proc. Soc. Exp. Biol. Med.* 36, 141–144.

Harne O. G., Lutz J. F., Zimmerman G. I., and Davis C. L. (1945). The life duration of the red blood cell of the *Macacus rhesus* monkey. *J. Lab. Clin. Med.* 30, 247–258.

Hartwig Q. L., Melville G. S., Leffingwell T. P., and Young R. J. (1958). Iron-59 metabolism as an index of hematopoietic damage and recovery in monkeys exposed to nuclear radiation. USAF School of Aviation Medicine. Report No 58–59, March, 1958. Cited by Krise G. M. and Wald N. 1959.

Hassimoto M., Harada T., and Harada T. (2004). Changes in hematology, biochemical values, and restraint ECG of rhesus monkeys (*Macaca mulatta*) following 6-month laboratory acclimation. *J. Med. Primatol.* 33, 175–186.

Hawkey C. M. (1991). The value of comparative haematological studies. *Comp. Haematol. Int.* 1, 1–9.

Hayes K. C. (1974). Pathophysiology of vitamin E deficiency in monkeys. *Am. J. Clin. Nutr.* 27, 1130–1140.

Hemm R. D. and Johnson N. W. (1978). Hematologic and serum chemistry values following phencyclidine administration to *Macaca mulatta. Toxicol. Appl. Pharmacol.* 43, 279–285.

Herndon J. G., Turner J. J., Perachio A. A., Blank M. S. et al. (1984). Endocrine changes induced by venipuncture in rhesus monkeys. *Physiol. Behav.* 32, 673–676.

Hillyer C. D., Brodie A. R., Ansari A. A., Anderson D. C., and McClure H. M. (1991). Severe autoimmune hemolytic anemia in SIV_{smm9}-infected *Macaca mulatta*. *J. Med. Primatol.* 20, 156–158.

Hillyer C. D., Duncan A., Ledford M., Barrett T. J. et al. (1995). Chemotherapy-induced hemolytic uremic syndrome: Description of a potential animal model. *J. Med. Primatol.* 24, 68–73.

Hillyer C. D., Klumpp S. A., Hall J. M., Lackey III D. A. et al. (1993). Multifactorial etiology of anemia in SIV-infected rhesus macaques: Decreased BFU-E formation, serologic evidence of autoimmune hemolysis, and an exuberant erythropoietin response. *J. Med. Primatol.* 22, 253–256.

Hom G. J., Bach T. J., Carroll D., Forrest M. J. et al. (1999). Comparison of cardiovascular parameters and/or serum chemistry and hematology profiles in conscious and anesthetized rhesus monkeys (*Macaca mulatta*). *Contemp. Top.* 38, 60–64.

Huser H.-J. (1970). *Atlas of Comparative Primate Hematology*. Academic Press, New York, pp. 158–178, 195–198.

Ibanez-Contreras A., Hernandez-Godinez B., Reyes-Pantoja S. A., Jimenez-Garcia A. et al. (2013). Changes in blood parameters in rhesus monkeys (*Macaca mulatta*) during the first trimester of gestation. *J. Med. Primatol.* 42, 171–176.

Ives M. and Dack G. M. (1956). "Alarm reaction" and normal blood picture in *Macaca mulatta*. *J. Lab. Clin. Med.* 47, 723–729.

Jagoe C. H. and Welter D. A. (1995). Quantitative comparisons of the morphology and ultrastructure of erythrocyte nuclei from seven freshwater fish species. *Can. J. Zool.* 73, 1951–1959.

Jones O. P. (1969). Elimination of midbodies from mitotic erythroblasts and their contribution to fetal blood plasma. *J. Nat. Cancer Inst.* 42, 753–763.

Kessler M. J. and Rawlins R. G. (1983). The hemogram, serum biochemistry, and electrolyte profile of the free-ranging Cayo Santiago rhesus macaques (*Macaca mulatta*). *Am. J. Primatol.* 4, 107–116.

Kessler M. J., Rawlins R. G., and London W. T. (1983). The hemogram, serum biochemistry, and electrolyte profile of aged rhesus monkeys (*Macaca mulatta*). *J. Med. Primatol.* 12, 184–191.

Kim C.-Y., Han J. S., Suzuki T., and Han S.-S. (2005). Indirect indicator of transport stress in hematological values in newly acquired cynomolgus monkeys. *J. Med. Primatol.* 34, 188–192.

Kim C.-Y., Lee H.-S., Han S.-C., Heo J.-D. et al. (2005). Hematological and serum biochemical values in cynomolgus monkeys anesthetized with ketamine hydrochloride. *J. Med. Primatol.* 34, 96–100.

King T. O. and Gargus J. L. (1967). Normal blood values of the adult female monkey (*Macaca mulatta*). *Lab. Anim. Care* 17, 391–396.

Kjeldsberg C., Beutler E., Bell C., Hougie C. et al. (1989). Chapter 3. Iron deficiency disease; Chapter 8. Hereditary erythrocyte membrane defects; Appendix. In: Kjeldsberg C. (ed.), *Practical Diagnosis of Hematologic Disorders*. American Society of Clinical Pathologists, Chicago, IL, pp. 31–41, 110–117, 632–636.

Klumpp S. A., Hillyer C. D., Olberding B. A., and McClure H. M. (1991). Significant bone marrow abnormalities in SIVsmm infected rhesus macaques: Analysis of aspirates and biopsies. In *Proceedings of the Ninth Annual Sympsium on Nonhuman Primate Models for AIDS*, Seattle, WA, p. 105 (abstract).

Kobayashi M., Asano H., and Hotta T. (1990). Occurrence of cytoplasmic bridge between erythroblasts—Morphology and frequency in hematological diseases. *Jpn. J. Clin. Hematol.* 31, 946–950.

Koga T., Kanefuji K., and Nakama K. (2005). Individual reference intervals of hematological and serum biochemical parameters in cynomolgus monkeys. *Int. J. Toxicol.* 24, 377–385.

Kratz A., Ferraro M., Sluss P. M., and Lewandrowski K. B. (2004). Laboratory reference values. *N. Engl. J. Med.* 351, 1548–1563.

Kreier J. P., Swann A. I., Taylor W. M., and Wagner W. M. (1970). Erythrocyte life span and label elution in monkeys (*Macaca mulatta*) and cats (*Felis catus*) determined with chromium-51 and diisopropyl fluorophosphate-32. *Am. J. Vet. Res.* 31, 1429–1435.

Kriete M. F., Champoux M., and Suomi S. J. (1992). Hematological parameters in mother-reared and nursery/peer-reared rhesus macaque infants. *Am. J. Primatol.* 27, 40 (abstract).

Krise G. M. and Wald N. (November 1959). Hematological effects of acute and chronic experimental blood loss in the *Macaca mulatta* monkey. *Am. J. Vet. Res.* 20, 1081–1085.

Krumbhaar E. B. and Musser J. H. Jr. (1921). Studies of the blood of normal monkeys. *J. Med. Res.* 42, 105–109.

Krumbhaar E. B. and Musser J. H. Jr. (1923). The effect of splenectomy on the hemopoietic system of *Macacus rhesus. Arch. Int. Med.* 31, 686–700.

Kuksova M. I. and Dikovenko E. A. (1961). The leucocyte reaction dynamics in lower monkeys. In: *Vop. Physiol. i Patologii Obesjan.* Sukhumi, Russia, pp. 25–34. Cited by Lapin and Cherkovich 1972.

Kupper J. L., Kessler M. J., and Cook L. L. (1976). Normal hematological, biochemical, and serum electrolyte values for a colony of rhesus monkeys (*Macaca mulatta*). Naval Aerospace Medical Research Report 1230, Naval Aerospace Medical Research Laboratory, Pensacola, FL.

Lajtha L. G. (1965). Cellular mechanism of red cell production. *Scand. J. Haematol. Ser. Haematol.* 2, 265–233.

Lapin B. A. and Cherkovich G. M. (1972). Biological normals. In: *Pathology of Simian Primates, Part I.* Karger, Basel, pp. 78–156.

Lee J.-I., Shin J.-S., Lee J.-E., Jung W.-Y. et al. (2013). Changes of N/L ratio and cortisol levels associated with experimental training in untrained rhesus macaques. *J. Med. Primatol.* 42, 10–14.

Letvin N. L. and King N. W. (1990). Immunologic and pathologic manifestations of the infection of rhesus monkeys with simian immunodeficiency virus of macaques. *J. Acquir. Immune Def. Syndr.* 3, 1023–1040.

Lewis J. H. (1977). Comparative hematology: Rhesus monkeys (*Macaca mulatta*). *Comp. Biochem. Physiol.* 56A, 379–383.

Li F., Lu S.-J., and Honig G. R. (2006). Hematopoietic cells from primate embryonic stem cells. *Methods Enzymol.* 418, 243–251.

Lilly A. A., Mehlman P. T., and Higley J. D. (1999). Trait-like immunological and hematological measures in female rhesus across varied environmental conditions. *Am. J. Primatol.* 48, 197–223.

Line S. W., Clarke A. S., and Markowitz H. (1987). Plasma cortisol of female rhesus monkeys in response to acute restraint. *Lab. Primate Newslett.* 26(4), 1–4.

Liu D. X., Gill A., Holman P. J., Didier P. J. et al. (2014). Persistent babesiosis in rhesus macaque (*Macaca mulatta*) infected with a simian-human immunodeficiency virus. *J. Med. Primatol.* 43, 206–208. doi: 10.1111/jmp.12105.

Loeb W. F., Bannerman R. M., Rininger B. F., and Johnson A. J. (1978). Hematologic disorders. In: Benirschke K., Garner F. M., and Jones T. C. (eds.), *Pathology of Laboratory Animals*, Vol. I, Chapter 11. Springer-Verlag, New York, pp. 890–1050.

Loomis L. J., Aronson A. J., Rudinsky R., and Spargo B. H. (1989). Hemolytic uremic syndrome following bone marrow transplantation: A case report and review of the literature. *Am. J. Kid. Dis.* XIV, 324–328.

Loomis M. R., Henrickson R. V., and Anderson J. H. (1980). Effects of ketamine hydrochloride on the hemogram of rhesus monkeys (*Macaca mulatta*). *Lab. Anim. Sci.* 30, 851–853.

Lowenstine L. J. and Lerche N. W. (1988). Retrovirus infections of nonhuman primates: A review. *J. Zoo Anim. Med.* 19, 168–187.

Lugo-Roman L. A., Rico P. J., Sturdivant R., Burks R. et al. (2010). Effects of serial anesthesia using ketamine or ketamine/medetomidine on hematology and serum biochemistry values in rhesus macaques (*Macaca mulatta*). *J. Med. Primatol.* 39, 41–49.

MacKenzie M., Lowenstine L., Lalchandani R., Lerche N. et al. (1986). Hematologic abnormalities in simian acquired immune deficiency syndrome. *Lab. Anim. Sci.* 36, 14–19.

Mandell C. P. and George J. W. (1991). Effect of repeated phlebotomy on iron status of rhesus monkeys (*Macaca mulatta*). *Am. J. Vet. Res.* 52, 728–733.

Mandell C. P., Jain N. C., Miller C. J., Marthas M. et al. (1992). Early events and cellular targets in bone marrow of SIV-infected rhesus macaques. In *Proceedings of the 10th Annual Symposium on Nonhuman Primate Models for AIDS*, San Juan, Puerto Rico (abstract 7).

Mandell C. P., Jain N. C., Miller C. J., Marthas M. et al. (1993). Early hematologic changes in rhesus macaques (*Macaca mulatta*) infected with pathogenic and nonpathogenic isolates of SIVmac. *J. Med. Primatol.* 22, 177–186.

Manwani D. and Bieker J. J. (2008). The erythroblastic island. *Curr. Top. Dev. Biol.* 82, 23–53.

Martin D. P., McGowan M. J., and Loeb W. F. (1973). Age related changes of hematologic values in infant *Macaca mulatta*. *Lab. Anim. Sci.* 23, 194–200.

Marvin H. N., Dinning J. S., and Day P. L. (1960). Erythrocyte survival in vitamin E-deficient monkeys. *Proc. Soc. Exp. Biol. Med.* 105, 473–475.

Mathis C. et Leger M. (1911). *Plasmodium* des macaques du Tonkin. *Ann. L'Institut Pasteur* 25, 593–600 avec Planche I.

Matsumoto K., Akagi H., Ochiai T., Hagino K. et al. (1980). Comparative blood values of *Macaca mulatta* and *Macaca fascicularis*. *Exp. Anim.* 29, 335–340.

Matsuzawa T. and Nagai Y. (1994). Comparative haematological and plasma chemistry values in purpose-bred squirrel, cynomolgus and rhesus monkeys. *Comp. Haematol. Int.* 4, 43–48.

Matsuzawa T., Nomura M., and Unno T. (1993). Clinical pathology reference ranges of laboratory animals. *J. Vet. Med. Sci.* 55, 351–362.

May C. D., Nelson E. N., and Salmon R. J. (1949). Experimental production of megaloblastic anemia; an interrelationship between ascorbic acid and pteroylglutamic acid. *J. Lab. Clin. Med.* 34, 1724–1725.

May C. D., Nelson E. N., Salmon R. J., Lowe C. U. et al. (1950a). Experimental production of megaloblastic anemia in relation to megaloblastic anemia in infants. *Bull. Univ. Minn. Minn. Med. Found.* 21, 208–222.

May C. D., Sundberg R. D., and Schaar F. (1950b). Comparison of effects of folic acid and folinic acid in experimental megaloblastic anemia. *J. Lab. Clin. Med.* 36, 963–964.

May C. D., Sundberg R. D., Schaar F., Lowe C. U. et al. (1951). Experimental nutritional megaloblastic anemia: Relation of ascorbic acid and pteroylglutamic acid. I. Nutritional data and manifestations of animals. *Am. J. Dis. Child.* 82, 282–309.

McCall K. B., Waisman H. A., Elvehjem C. A., and Jones E. S. (1946). A study of pyridoxine and pantothenic acid deficiencies in the monkey (*Macaca mulatta*). *J. Nutr.* 31, 685–697.

Merritt C. B. and Gengozian N. (1967). Survival of ^{51}Cr labeled cells in marmosets, U.S. Atomic Energy Commission Report ORAU-106, Washington, DC, pp. 208–211.

Mohandas N. and Prenant M. (1978). Three-dimensional model of bone marrow. *Blood* 51, 633–643.

Monroy R. L., MacVittie T. J., Darden J. H., Schwartz G. N. et al. (1986). The rhesus monkey: A primate model for hemopoietic stem cell studies. *Exp. Hematol.* 14, 904–911.

Morrow A. C. and Terry M. W. (1970a). *Hematologic Values for Nonhuman Primates Tabulated from the Literature. I. Erythrocytes, Hemoglobin, Hematocrit and Related Indexes.* Primate Information Center, Regional Primate Research Center, University of Washington, Seattle, WA, 46pp.

Morrow A. C. and Terry M. W. (1970b). *Hematologic Values for Nonhuman Primates Tabulated from the Literature. II. Total Leucocytes and Differential Counts.* Primate Information Center, Regional Primate Research Center, University of Washington, Seattle, WA, 43pp.

Munro N. (1987). A three year study of iron deficiency and behavior in rhesus monkeys. *Int. J. Biosoc. Res.* 9, 35–62.

National Research Council. (1998). Effect of special research conditions on psychological well-being calorie restriction study in rhesus monkeys. *Exp. Gerontol.* 33, 421–443. In: *The Psychological Well-Being of Nonhuman Primates*, Chapter 4. National Academy Press, Washington, DC, 54pp.

Nirmalan G. P. and Robinson G. A. (1971). Haematology of the Japanese quail (*Coturnix coturnix japonica*). *Br. Poult. Sci.* 12, 475–481.

O'Sullivan M. G., Anderson D. C., Fikes J. D., Bain F. T. et al. (1994). Identification of a novel simian parvovirus in cynomolgus monkeys with severe anemia. *J. Clin. Invest.* 93, 1571–1576.

O'Sullivan M. G., Anderson D. K., Lund J. E., Brown W. P. et al. (1996). Clinical and epidemiological features of simian parvovirus infection in cynomolgus macaques with severe anemia. *Lab. Anim. Sci.* 46, 291–297.

Overman R. R. and Feldman H. A. (1947). Circulatory and fluid compartment physiology in the normal monkey with especial reference to seasonal variations. *Am. J. Physiol.* 148, 455–459.

Perretta G., Violante A., Scarpulla M., Beciani M. et al. (1991). Normal serum biochemical and hematological parameters in *Macaca fascicularis. J. Med. Primatol.* 20, 345–351.

Ponder E., Yeager J. F., and Charipper H. A. (1928a). Studies in comparative haematology: I. Camelidae. *Quart. J. Exp. Physiol.* 19, 115–126.

Ponder E., Yeager J. F., and Charipper H. A. (1928b). Studies in comparative haematology: II. Primates. *Quart. J. Exp. Physiol.* 19, 181–195.

Ponder E., Yeager J. F., and Charipper H. A. (1929). Haematology of the primates. *Zoologica* 11, 9–18.

Poppen K. J., Greenberg L. D., and Rinehart J. F. (1952). The blood picture of pyridoxine deficiency in the monkey. *Blood* 7, 436–494.

Porter F. S., Fitch C. D., and Dinning J. S. (1962). Vitamin E deficiency in the monkey. IV. Further studies of the anemia with emphasis on bone marrow morphology. *Blood* 20, 471–477.

Premasuthan A., Ng J., Kanthaswamy S., Satkoski J. et al. (2012). Molecular ABO phenotyping in cynomolgus macaques using real time quantitative PCR (QPCR). *Tissue Antigens* 80, 363–367.

Puri, C. P., Puri V., and Anand Kumar T. C. (1981). Serum levels of testosterone, cortisol, prolactin and bioactive luteinizing hormone in adult male rhesus monkeys following cage-restraint or anesthetizing with ketamine hydrochloride. *Acta Endocrinol.* 97, 118–124.

Qiong C., Huayou Z., Zepeng L., Junhua R., et al. (2003). Report concerning ABO blood groups in Chinese. *Chinese J. Comp. Med.* 13 (2), 94–95 (abstract). [Title not presented in English abstract. Site of study: Department of Blood Transfusion, Nanfang Hospital. Guangzhou 510515, Guangdong, China.]

Rabinowe S. N., Soiffer R. J., Tarbell N. J., Neuberg D. et al. (1991). Hemolytic-uremic syndrome following bone marrow transplantation in adults for hematologic malignancies. *Blood* 77, 1837–1844.

Rakieten N. (1935). The basal heat production of the rhesus monkey (*Macaca mulatta*). *J. Nutr.* 10, 357–362.

Reinhardt V. (1991a). Training adult male rhesus monkeys to actively cooperate during in-homecage venipuncture. *Anim. Technol.* 42, 11–17.

Reinhardt V. (1991b). Impact of venipuncture on physiological research conducted in conscious macaques. *J. Anim. Sci.* 34, 212–217.

Reinhardt V. (1992a). Improved handling of experimental rhesus monkeys. In: Davis H. and Balfour A. D (eds), *The Inevitable Bond: Examining Scientist-Animal Interactions*, Chapter 10. Cambridge University Press, Cambridge, UK, pp. 171–177.

Reinhardt V. (1992b). Difficulty in training juvenile rhesus macaques to actively cooperate during venipuncture in the homecage. *Lab. Primate Newslett.* 31(3), 1–2.

Reinhardt V. (1996). Refining the blood collection procedure for macaques. *Lab Anim.* 1, 32–35.

Reinhardt V. (2003). Working with rather than against macaques during blood colection. *J. Appl. Anim. Welfare Sci.* 6, 189–197.

Reinhardt V., Cowley D., Eisele S., and Scheffler J. (1991). Avoiding undue cortisol responses to venipuncture in adult male rhesus macaques. *Anim. Technol.* 42, 83–86.

Reinhardt V., Cowley D., Scheffler J., Vertein R. et al. (1990). Cortisol response of female rhesus monkeys to venipuncture in homecage versus venipuncture in restraint apparatus. *J. Med.* 19, 601–606.

Reinhardt V. and Reinhardt A. (2001). *Environmental Enrichment for Caged Rhesus Macaques*, 2nd edn. Animal Welfare Institute, Washington, DC, 77pp.

Ridgway S. H. (1972). Homeostasis in the aquatic environment. In: Ridgway S. H. (ed), *Mammals of the Sea: Biology and Medicine*, Chapter 10. Charles C. Thomas Publisher, Springfield, IL, pp. 590–747.

Robinson F. R. and Ziegler R. F. (1968). Clinical laboratory data derived from 102 *Macaca mulatta. Lab. Anim. Care* 18, 50–57.

Rodnan G. P., Ebaugh F. G., Jr., and Spivey Fox M. R. (1957). The life span of the red blood cell and the red blood cell volume in the chicken, pigeon, and duck as estimated by the use of $Na_2Cr^{51}O_4$. With observations on red cell turnover rate in the mammal, bird and reptile. *Blood* 12, 355–366.

Rogers L. B., Kaack M. B., Henson M. C., Rasmussen T. et al. (2005). Hematologic and lymphocyte immunophenotypic reference values for normal rhesus monkey (*Macaca mulatta*) umbilical cord blood; gravidity may play a role in study design. *J. Med. Primatol.* 34, 147–153.

Roney E. E. (1971). Use of the blowgun to immobilize or medicate caged animals. *J. Zoo Anim. Med.* 2(2), 25.

Rosenblum I. Y. and Coulston F. (1981). Normal range and longitudinal blood chemistry and hematology values in juvenile and adult rhesus monkeys (*Macaca mulatta*). *Ecotoxicol. Environ. Saf.* 5, 401–411.

Rosenzweig M., Marks D. F., DeMaria M. A., Connole M. et al. (2001). Identification of primitive hematopoietic progenitor cells in the rhesus macaque. *J. Med. Primatol.* 30, 36–45.

Rowe A. W. and Davis J. H. (1972). Erythrocyte survival in chimpanzees, gibbons and baboons. *J. Med. Primatol.* 1, 86–89.

Ruebush T. K. II, Collins W. E., and Warren M. (1981). Experimental *Babesia microti* infections in *Macaca mulatta*: Recurrent parasitemia before and after splenectomy. *Am. J. Trop. Med. Hyg.* 30, 304–307.

Sae-Low W. and Malaivijitnond S. (2003). The determination of human-ABO blood groups in captive cynomolgus macaques (*Macaca fascicularis*). *Nat. Hist. J. Chulalongkorn Univ.* 3(1), 55–60.

Santiyanont R., Yaipimol C., and Wilairat P. (1977). Accumulation of orthochromatophilic normoblasts in bone marrow of vitamin E-deficient monkey, *Macaca fascicularis*. *J. Nutr.* 107, 2026–2030.

Schmidt L. H., Greenland R., Rossan R., and Genther C. (1961). Natural occurrence of malaria in rhesus monkeys. *Science* 133, 753.

Schuurman H.-J. and Smith H. T. (2005). Reference values for clinical chemistry and clinical hematology parameters in cynomolgus monkeys. *Xenotransplantation* 12, 72–75.

Selye H. (1936). A syndrome produced by diverse nocous agents. *Nature* 138, 32.

Smucny D. A., Allison D. B., Ingram D. K., Roth G. S. et al. (2001). Changes in blood chemistry and hematology variables during aging in captive rhesus macaques (*Macaca mulatta*). *J. Med. Primatol.* 30, 161–173.

Smucny D. A., Allison D. B., Ingram D. K., Roth G. S. et al. (2004). Update. Changes in blood chemistry and hematology variables during aging in captive rhesus macaques (*Macaca mulatta*). *J. Med. Primatol.* 33, 48–54.

Sood S. K., Deo M. G., and Ramalingaswami V. (1965). Anemia in experimental protein deficiency in the rhesus monkey with special reference to iron metabolism. *Blood* 26, 421–432.

Socolowsky M. (2013). Exploring the erythroblastic island. *Nat. Med.* 19, 399–401.

Sorrell J. M. and Weiss L. (1982). Intercellular junctions in the hematopoietic compartments of embryonic chick bone marrow. *Am. J. Anat.* 164, 57–66.

Spicer E. J. F. and Oxnard C. E. (1967). Some haematological changes during pregnancy in the rhesus monkey (*Macaca mulatta*). *Folia Primatol.* 6, 236–242.

Stanley R. E. and Cramer M. B. (1968). Hematologic values of the monkey (*Macaca mulatta*). *Am. J. Vet. Res.* 29, 1041–1047.

Stasney J. and Higgins G. M. (1936). The bone marrow in the monkey (*Macacus rhesus*). *Anat. Rec.* 67, 219–231.

Suarez R. M., Diaz Rivera R. S., and Hernandez Morales F. (1942). Hematological studies in normal rhesus monkeys (*Macaca mulatta*). *Puerto Rico J. Pub. Health Trop. Med.* 18, 212–226.

Suarez R. M., Diaz-Rivera R. S., and Hernandez-Morales F. (1943). Aspirated bone marrow studies in normal *Macacus rhesus* monkeys. *Am. J. Med. Sci.* 205, 581–586.

Sugimoto Y., Hanari K., Narita H., and Honjo S. (1986a). Normal hematologic values in the cynomolgus monkeys aged from 1 to 18 years. *Exp. Anim.* 35, 443–447.

Sugimoto Y., Ohkubo F., Ohtoh H., and Honjo S. (1986b). Changes of hematologic values for 11 months after birth in the cynomolgus monkeys. *Exp. Anim.* 35, 449–454.

Sundberg R. D., Schaar F., and May C. D. (1952). Experimental nutritional megaloblastic anemia. II. Hematology. *Blood* 7, 1143–1181.

Suzuki J., Gotoh S., Miwa N., Terao K. et al. (2000). Autoimmune hemolytic anemia (AIHA) in an infant rhesus macaque (*Macaca mulatta*). *J. Med. Primatol.* 29, 88–94.

Switzer J. W. (1967a). Bone marrow composition in the adult rhesus monkey (*Macaca mulatta*). *J. Am. Vet. Med. Assoc.* 151, 823–829.

Switzer J. W. (1967b). A new technique for sampling bone marrow in monkeys. *Lab. Anim. Care* 17, 255–260.

Switzer J. W., Valerio D. A., Martin D. P., Valerio M. G. et al. (1970). Hematologic changes associated with pregnancy and parturition in *Macaca mulatta*. *Lab. Anim. Care* 20, 930–939.

Takenaka O. (1981). Blood characteristics of the crab-eating monkeys (*Macaca fascicularis*) in Bali and Sumatra. *Kyoto Univ. Overseas Rep. Stud. Indones. Macaque* 1, 41–46.

Takenaka O. (1986). Blood characteristics of the crab-eating monkeys (*Macaca fascicularis*) in Bali Island, Indonesia: Implication of water deficiency in West Bali. *J. Med. Primatol.* 15, 97–104.

Taketa S. T., Carsten A. L., Cohn S. H., Atkins H. L. et al. (1970). Active bone marrow distribution in the monkey. *Life Sci.* 9(Part II), 169–174.

Tarantal A. F. (1993). Hematologic reference values for the fetal long-tailed macaque *Macaca fascicularis*. *Am. J. Primatol.* 29, 209–219.

Tarantal A. F., Goldstein O., Barley F., and Cowan M. J. (2000). Transplantation of human peripheral blood stem cells into fetal rhesus monkeys (*Macaca mulatta*). *Transplantation* 69, 1818–1823.

Terao K. (2005). Management of old world primates. In: Wolfe-Coote S. (ed.), *The Laboratory Primate*, Chapter 11. A volume of The Laboratory Animal Series, *Handbook of Experimental Animals*, Petrusz P. and Bullock G. (eds.-in-chief). Elsevier, Boston, MA, pp. 163–173.

Thiebot H., Louache F., Vaslin B., de Revel T. et al. (2001). Early and persistent bone marrow hematopoiesis defect in simian/human immunodeficiency virus-infected macaques despite efficient reduction of viremia by highly active antiretroviral therapy during primary infection. *J. Virol.* 75, 11594–11602.

Toda S., Segawa K., and Nagata S. (2014). MerTK-mediated engulfment of pyrenocytes by central macrophages in erythroblastic islands. *Blood* 123, 3963–3971.

Umeda K., Heike T., Yoshimoto M., Shiota M. et al. (2004). Development of primitive and definitive hematopoiesis from non-human primate embryonic stem cells in vitro. *Development* 131, 1869–1879.

Uno H. (1997). Age-related pathology and biosenescent markers in captive rhesus macaques. *Age* 20, 1–13.

Usacheva I. N. and Raeva N. V. (1963). Normal indices of the peripheral blood and bone marrow for the monkey *Macacus rhesus*. *Bull. Exp. Biol. Med.* 54, 1285–1287. Translated from *Byulleten Eksperimental noi Biologii i Meditsiny* 54(11), 106–108, November 1962.

Vacha J. (1983). Red cell life span. In: Agar N. S. and Board P. G. (eds.), *Red Blood Cells of Domestic Mammals*, Chapter 4. Elsvevier, Amsterdam, the Netherlands, pp. 67–132.

Van Beusechem V. W. and Valerio D. (1996). Gene transfer into hematopoietic stem cells of nonhuman primates. *Hum. Gene Ther.* 7, 1649–1668.

Verlangieri A. J., DePriest J. C., and Kapeghian J. C. (1985). Normal serum biochemical, hematological and EKG parameters in anesthetized adult male *Macaca fascicularis* and *Macaca arctoides*. *Lab. Anim. Sci.* 35, 63–66.

Vertein R. and Reinhardt V. (1989). Training female rhesus monkeys to cooperate during in-homecage venipuncture. *Lab. Primate Newslett.* 28(2), 1–3.

Vondruska J. F. and Greco R. A. (1974). Certain hematologic and blood chemical values in juvenile rhesus monkeys (*Macaca mulatta*). *Bull. Am. Soc. Vet. Clin. Pathol.* 3, 27–36.

Walker M. L. (1995). Menopause in female rhesus monkeys. *Am. J. Primatol.* 35, 59–71.

Wang H., Niu Y. Y., Si W., and Yan Y. (2012). Reference data of clinical chemistry, haematology and blood coagulation parameters in juvenile cynomolgus monkeys (*Macaca fascicularis*). *Vet. Med.* 57, 233–238.

Watanabe M., Ringler D. J., Nakamura M., DeLong P. A. et al. (1990). Simian immunodeficiency virus inhibits bone marrow hematopoietic progenitor cell growth. *J. Virol.* 64, 656–663.

Wel A. V., Kocken C. H. M., Zeeman A.-M., and Thomas A. W. (2008). Short report: Detection of new *Babesia microti*-like parasites in a rhesus monkey (*Macaca mulatta*) with a suppressed *Plasmodium cynomolgi* infection. *Am. J. Trop. Med. Hyg.*78, 643–645.

Wickings E. J. and Nieschlag E. (1980). Pituitary response to LRH and TRH stimulation and peripheral steroid hormones in conscious and anesthetized adult male rhesus monkeys (*Macaca mulatta*). *Acta Endocrinol.* 93, 287–293.

Wills L. and Bilimoria H. S. (1932). Studies in pernicious anaemia of pregnancy. Part V. Production of a macrocytic anaemia in monkeys by deficient feeding. *Indian J. Med. Res.* 20, 391–402.

Wills L. and Stewart A. (1935). Experimental anaemia in monkeys with special reference to macrocytic nutritional anaemia. *Br. J. Exp. Pathol.* 16, 444–453.

Wilson W. O., Abbott U. K., and Abplanalp H. (1961). Evaluation of *Coturnix* (Japanese quail) as pilot animal for poultry. *Poult. Sci.* 40, 651–657.

Winkle V. A. (1951). The study of bone marrow and blood of normal laboratory monkeys. Thesis, University of Minnesota, Minneaolis, MN.

Winterborn A. N., Bates W. A., Feng C., and Wyatt J. D. (2008). The efficacy of orally dosed ketamine and ketamine/medetomidine compared with intramuscular ketamine in rhesus macaques (*Macaca mulatta*) and the effects of dosing route on haematological stress markers. *J. Med. Primatol.* 37, 116–127.

Wintrobe M. M. (1933). Variations in the size and hemoglobin content of erythrocytes in the blood of various vertebrates. *Folia Haematol.* 51, 32–39.

Wintrobe M. M., Lee G. R., Boggs D. R., Bithell T. C. et al. (1974). Destruction of erythrocytes. The hemolytic disorders: General considerations. Normal values for osmotic fragility. In: *Clinical Hematology*, 7th edn, Chapters 5, 20, and Appendix A-9. Lea and Febiger, Philadelphia, PA, pp. 195–220, 717–750, and Table A-9 on page 1793.

Wixson S. K. and Griffith J. W. (1986). Nutritional deficiency anemias in nonhuman primates. *Lab. Anim. Sci.* 36, 231–236.

Wolcott G. J., Valentine J. A., and Cebul R. D. (1973). Induction of anaemia in monkeys. *Lab. Anim.* 7, 297–303.

Wolford S. T., Schroer R. A., Gohs F. X., Gallo P. P. et al. (1986). Reference range data base for serum chemistry and hematology values in laboratory animals. *J. Toxicol. Environ. Health* 18, 161–188.

Woodward R. A. and Weld K. P. (1997). A comparison of ketamine, ketamine-acepromazine, and tiletamine-zolazepam on various hematologic parameters in rhesus monkeys (*Macaca mulatta*). *J. Am. Assoc. Lab. Anim. Sci.* 36(3), 55–57.

Xie L., Xu F., Liu S., Ji Y. et al. (2013). Age- and sex-based hematological and biochemical parameters for *Macaca fascicularis*. *PLos One* 8(6), e64892. Published online June 10, 2013. doi: 10. 1371/journal.pone.0064892.

Yoshida T. (1981). The changes of hematological and biochemical properties in cynomolgus monkeys (*Macaca fascicularis*) after importation. *Jpn. J. Med. Sci. Biol.* 34, 239–242.

Yoshida T., Katsuta A., and Cho F. (1989). Reference values of hematological, serum bio-chemical and hormonal examinations in female cynomolgous monkeys (*Macaca fascicularis*) of feral origin. *Exp. Anim.* 38, 259–262.

Yoshida T., Ohtoh K., Cho F., Honjo S. et al. (1988). Discriminant analyses for pregnancy-related changes in hematological and serum biochemical values in cynomolgus mon-keys (*Macaca fascicularis*). *Jikken Dobutsu* 37, 257–262. Abstract in English.

Yoshida T., Suzuki K., Cho F., and Honjo S. (1986a). Age-related changes of hematologi-cal and serum biochemical values in cynomolgus monkeys (*Macaca fascicularis*) bred and reared using the indoor individually-caged system. *Exp. Anim.* 35, 329–338.

Yoshida T., Suzuki K., Shimizu T., Cho F. et al. (1986b). The effects of ketamine anesthe-sia on hematological and serum biochemical values in female cynomolgus monkeys (*Macaca fascicularis*). *Exp. Anim.* 35, 455–461.

Zeng X.-C., Yang C.-M., Pan X.-Y., Yao Y.-S. et al. (2011). Effects of fasting on hematologic and clinical chemical values in cynomolgus monkeys (*Macaca fascicularis*). *J. Med. Primatol.* 40, 21–26.

Zon L. I., Arkin C., and Groopman J. E. (1987). Haematologic manifestations of the human immune deficiency virus (HIV). *Br. J. Haematol.* 66, 251–256.

Zon L. I. and Groopman J. E. (1988). Hematologic manifestations of the human immune deficiency virus (HIV). *Sem. Hematol.* 25, 208–218.

Index